# Howell Davidson

the

# Musketeer

Being a True and Accurate Account of the Adventure of

# Howell Davidson

the

# Musketeer

by

Steven R. Culver

H&E
Harnew &
Edgebrook
Publishing

**ISBN: 978-0-9886437-4-1**



**Book Cover Design by eBookLaunch.com**

# Thank You

I owe thanks to many people for what this book has become. Susan has been an invaluable critic, and has helped me see things from a new perspective any number of times. Laurie and London have been my writing buddies for years, and have supported my efforts all along. Not the least of my thanks must go to my wife, who has endured endless hours of my typing while my music played 'way too loud.

**My thanks to London for the intra-chapter divider.**

**and the Publisher logo**

In Memory of

# John Delmerico

# Cameraman

and

# Wit Extraordinaire

# Chapter I

## Pirates of the Mediterranean

Howell Davidson stood at the ship's rail under a beautiful cloudless sky, looking out over the calm Mediterranean Sea rolling by, and knew that before this day was done, he would be fighting for his life. The lookout on the mast had been the first to spy the pirate, and the first to announce the terrible news. The sail was far out on the horizon, but Howell could see it now and then as the waves swelled, and raised it into view.

"She's lanteen rigged, Captain," the lookout called down, "and she's closing fast."

The Captain had his eye to his telescope, studying what his man had pointed out. He gave all his attention to what he saw until finally he lowered his glass. "It is a pirate, and no mistake."

This confirmed not only what the man on the mast had seen, but it confirmed Howell's expectation of a battle.

"Excuse me Mr. Davidson, but could you explain to me what is happening?"

Howell shifted his attention from the pirate ship to the attractive woman who addressed him. She was about his age, say eighteen years. Her hair was dressed without vanity, but was attractive all the same. What was striking about her hair was that it was the purest red that Howell had ever seen. Her face was sprinkled with freckles. Altogether, a captivating creature. He had bantered with her since they began their voyage from London. She had tried to maintain her distance and to resist the temptation that he represented, but she had long ago submitted to his small talk.

"It would be my pleasure to tell you what I know, Miss Kincade." Howell had kept up appearances to the extent of using her proper name when he spoke aloud, but in his own mind he knew her by the nickname he had given her, Rosy. He had come to know her sufficiently well that this felt completely natural. "It would seem that we are being approached by a pirate."

"A pirate?"

"A pirate ship, yes."

She stood quietly for a moment, taking it in. "You are sure about this?"

"Ah yes, quite sure." Howell drew Rosy's attention to the place at the rail of the quarterdeck where the Captain stood, once again aiming his telescope where his lookout had pointed.

The Captain studied what he saw very carefully. Then he straightened up and snapped closed his telescope. "Two sails," he said to no one in particular.

A gentleman named Sir Horace now stepped up beside the Captain. "What's all this about pirates?"

Howell had found this man to be civil enough, but he had said very little during the voyage. About all that Howell knew about him was that a part of the ship's cargo was in his care, and that he saw it as his duty to make sure that it should be delivered safely to Genoa. He was a bit older than Howell, and sported a mane of blonde curls emerging from beneath his hat. He had two men with him who seemed to be on the ship to guard his cargo.

"I have heard of pirates in the Caribbean Sea, Captain, but these cannot be they," said Sir Horace.

"No, Sir, they are not. These are from Barbary."

"Barbary?"

The Captain turned to Sir Horace. "Barbary, Sir, from the north coast of Africa." The Captain pointed out across the sea. "Do you see those little triangles of white upon the horizon, fixed there while the waves move between us?"

Although the comment was not addressed to him, Howell knew what was meant, and pointed it out to Rosy. She nodded.

Sir Horace continued his investigation. "That is their ship, then?"

"Indeed."

"I can see now that their sails are very different from ours. Why do they sport such a peculiar arrangement?"

The Captain turned his telescope upon the threat once again. "It means that we are rigged to carry great deals of cargo, while they are rigged for speed that they may steal it." The Captain never took his eye off his telescope as he said this.

The Captain's explanation seemed to be sufficient to satisfy Sir Horace's curiosity. He took a few steps back from the Captain, and turned his gaze onto the horizon. He unconsciously put his hand onto the hilt of his sword.

Howell had his own reaction to consider. He felt an odd sensation about the prospect of fighting. It was a sort of enthusiasm, or anticipation. It was a feeling that he had not experienced in many years. It was a feeling that had ruled his life when he was an orphan in London's alleys, scrabbling for a living. He had fought the genteel who had spat upon him in the streets, and he had fought in the dark corners with the members of his own gang for his share of the loot. By great good fortune, he had been rescued from that world. While he had labored to drag himself from scullion to clerk in a counting house, his warrior feeling had faded away. No longer. It was back, unbidden, rising, and seizing his heart.

"A pirate ship," said Rosy. "I shall have quite a story to tell Elizabeth when next we meet."

"Amen," said Howell. This was no longer small talk. "I feel confident that you will never forget what you see today, Miss. I fear it will mark you for life. The Countess will be thrilled to hear of your adventure, but you will never tell her how truly grim it was."

Rosy was sobered by this. She looked once more across the swell at the approaching ship. This would be serious business.

Howell was beginning to see enough of the pursuing ship to understand what the Captain had meant about the differences in their rigging. Howell pointed up their mast to draw Rosy's attention. "Look there. The sails above us are squares, more or less..." He turned to point at the vessel bearing down upon them. "…while that one has a very long yard at the top of each mast with a big sail like a triangle hanging upon it." Howell was amazed at what he saw of the pirate ship. "It appears that it can almost sail up wind." That boded ill. There was no possibility of outrunning nor outmaneuvering such a nimble vessel. He glanced at Rosy. Was that look upon her face wariness, or curiosity, or defiance?

Miss Kincade looked out at the pirate. "It seems to gain speed even as I watch."

The Captain stood at the rail, his eye pressed against his glass. He suddenly blurted out, "Surrender means slavery." He was speaking to himself. "They'll take us all to the slave market in Algiers. I will be a slave." He was terrified. He was paralyzed by the fate that loomed over them all. "If we fight, they will kill us. If we do not, we will all be slaves."

Howell took the Captain by his shoulders, whipped him around, and shook him violently. "Wake up, man! We won't be killed, and we won't be slaves if we defend the ship."

The Captain blinked his eyes. He looked at Howell without understanding what was being said to him.

"Command," said Howell. "Give your men the orders they need to defend themselves." He steered the Captain to the rail of his quarter deck.

The Captain looked down upon his crew. They were motionless, befuddled, and awaiting his orders. The Captain was not sure of what orders to give. "Arm yourselves!" he yelled to his crew and quickly glanced at the pirate ship that grew ominously closer every moment. "Arm yourselves!" He could think of nothing else to say.

The men saluted dutifully. "Aye, aye, Sir." However, none of these men, from the Captain to the cabin boy, had ever been in a situation like this. They had no idea what they should be doing, so they just stood and looked at one another.

Howell could see that far more needed to be done in what little time was left to them. He waited a long moment for something to happen, but nothing did. "Clear the decks! Remove anything that may hinder our defense."

The crew looked at their inert Captain. They looked at the firebrand beside their commander. They needed instructions, and unconsciously began to obey the confident man on the quarter deck above them.

Howell assumed the role of leader without authorization nor hesitation. He never paused to consider that he would be giving orders to men twice his age. "Clear the decks!"

The First Mate saluted. "Aye, aye, Sir." He tried to do his duty, but he had never done anything remotely like this before. "Stow those lines." His crew began to coil the ropes on the deck as they normally did. "You there, get those buckets below." Sailors descended into the hold to take the buckets that were handed down to them, and put them in their proper places. Most of the sailors were dealing with their ship in the orderly way that they had been taught. If anything, they were being particularly neat. No one understood what they were meant to do beyond their regular duties.

On the other side of the deck from Howell, Sir Horace shook his blonde curls as he bellowed, "Fools! If we let the ship be taken, it will all be for naught."

"Speed, sailors! Speed!" Howell slid down the stairs, his hands on the rails and his feet never touching the steps. "We must be ready before those cutthroats are upon us." He strode across the main deck, pointing at obstructions. He grabbed a bucket, and tossed it heedlessly into an open hatch. "In! In! Just throw it in! All of it!"

The crew scrambled to obey. They threw in mops, ropes, marlin spikes…

"Not that!"

One of the crew had attempted to toss in a belaying pin, a thick, foot-long rod used for tying ropes to the ship. Howell snatched it from the man's hand and held it high over his head. "Use these for clubs! Get out your knives! Arm yourself with whatever is to hand."

Sir Horace jumped up onto a hatch cover. He drew his rapier, and swept it to point along the side which the pirates would use for their attack. "There. Stand there. Spread out." Sir Horace then directed his commands to the two guards that he had personally added to the crew. "Spread them out. Put a knife between each of the clubs."

Horace's guards walked down the rough line of men standing at the rail with their improvised weapons. They shoved the sailors into position without ceremony. Howell agreed with Sir Horace that alternating clubs and knives would give them their best chance to drive back the pirates' attack.

Once again, Howell was facing the sort of challenge that he had faced so many years ago when he had fought for his life in the alleys. It was his gang versus the world. He had stood shoulder to shoulder with other boys as desperate and determined as he was. His lust for battle that he had held down for so long while he was in the business world had now returned, and was staring at him from across less than a mile of sea water.

Now. The men he led needed the resolve for the fight. The resolve that would give them victory. He worked his way down the back of the line, his six-foot frame towering over the sailors. He wanted them to feel that they could be as powerful as he was. He made necessary corrections and encouraged the men. "It will be just like killing a thief that springs upon you from an alley." Howell shook the man playfully. "Only this time you will be sober."

There was nothing playful about Sir Horace's harangue. "Strike while they are still coming up the side, do not wait for them to board." He stalked back and forth across his hatch cover shaking his sword. "Brain them. Slash them. Send these bastards so deeply into Hell that even their ghosts cannot escape. Keep these Devils from boarding altogether."

Howell had to agree. Hand-to-hand combat once the pirates were on deck would almost certainly be suicidal. Most of the crew had doubtless experienced more than a few drunken bar brawls, but this had certainly not made them hardened raiders like those that bore down upon them now.

He walked back down the line that he and Horace's toughs had sorted, and found himself at the bottom of the stairs that led up to the quarter deck.

The Mate looked stupidly at Howell.

"There!" Howell pointed up at the quarter deck. The ship had one distinct advantage. It carried a defense that the pirates would not expect. At both the forward corners of the quarter deck was fixed a small cannon called a 'murderer'. This was mounted on a swivel so that it could be instantly aimed in any direction. It did not generally fire solid shot like a larger, conventional cannon, but more often a handful of musket balls or any other jagged bits of metal that might come to hand. "Man that gun there! Sweep their decks before they even arrive."

The Mate came out of his stupor. The man he saw before him was in the dress of an office worker, yet this man was giving him sound advice. The Mate could never have imagined the circumstances that had spawned Howell, nor what Howell was capable of doing. A good course of action was good no matter where it came from, so the First Mate took one of his men. "You, come with me." They went up to man the murderer on the side on which the pirates were coming. "You are going to load this bloody thing." He looked at the approaching pirate. "And I am going to make bloody shreds out of those thieving whore's arses."

Their pursuer was close enough now that Howell could make out the bulging sails and maze of ropes that made their vessel so swift. He could see individual men moving about. He caught the glint of the Sun's light off the pirates' curving swords. A man at the stern ran up a black flag covered with some sort of complex, curling script that Howell had never seen before. A pirate near the bow cupped his hands around his mouth to make a sort of speaking trumpet, and called out, "Surrender! Surrender or die!" His accent was peculiar, but his meaning was plain enough.

"Come and get us, you bastards!" Howell's shout caught all on his ship by surprise. The men that surrounded him had been uncertain, even despondent, but now they took heart. Some of them smiled. Others chucked their mates in the arm.

Once again, Howell walked along the line, slapping the crewmen on the back and stiffening their spines. "They are men just like we are, lads. They shitte just like we do." The crew was beginning to believe that they might survive this battle after all.

Sir Horace was not smiling. He moved from one corner of his hatch cover to the other, over and over. "Steady. Focus."

The pirate ship was now close enough that one could throw a rock, and hit someone. Howell scanned the grinning faces below him as their ship drew nigh. All of them were swarthy from a life in the Sun. One was completely black. Many of them grinned at the anticipation of what was to come.

*Boom!* The murderer thundered, filling the air with gritty, grey-white smoke. Yelps and cries came from the pirate ship. Men scrambled about on her deck, but the vessel continued to close.

The attackers had not only their razor-sharp scimitars, but muskets as well, and these returned fire. Their particular aim was at the men serving the murderer, and the First Mate and his loader went down, screaming and groaning. The defenders' most potent weapon fell silent.

The pirate ship bumped into the side of the merchantman. The boarding party leapt up, and got their hands onto anything that they could grab. They pulled themselves up, and climbed the side. Sailors' clubs swung all along the side of the ship. Others slashed with their knives. Pirates attempted to get their swords where they could use them. The pirate's muskets cracked from their ship. The yelling grew louder and louder.

A pirate appeared directly before Howell, ready to come over the side and onto the deck. He carried a scimitar like most of the others, and the sweat ran down his brow. He looked more like a laborer doing a day's work than some kind of sea monster. Howell seized the wrist of the hand that held the sword and punched the man full in the face.

Blood from the pirate's smashed nose ran down over his mouth and chin. He looked angrily at Howell. He swung his free arm and fetched Howell a goodly blow on the side of his head.

This only increased Howell's fury. He twisted the man's wrist, and slammed the pirate's hand into the ship's rail. The pirate pulled back from Howell's hand, trying to free himself from Howell's crushing grip. Howell continued to pound the man's hand into the rail. He continued to punch his opponent's head. Now fully in the face, then on the side. Howell felt he was back in the streets, fighting for his life. In his mind, he was once again the friendless urchin in London's alleys. The pirate in his grip went slack. He dropped his sword, and it clattered onto the deck at Howell's feet. His enemy dangled briefly in Howell's grasp, and then Howell dropped him into the sea.

Howell snatched up the pirate's scimitar. He hefted it. He had wielded a knife many times, but this weapon was very different. It was much heavier and much longer. It was not meant to be used for stabbing like a knife, but it had a long, curved cutting edge that he could use for slashing his opponent. He felt that he knew exactly how to use it, even though he had never carried one before. It was going to man-to-man, and he found himself relishing the prospect.

A pirate leapt over the rail between Howell and the mass of struggling crewmen. Without a moment's hesitation, Howell chopped down on the man with all the strength in his arm. The pirate shrieked in pain, clutching his hand to the place where Howell had cut his back. He flew ahead to land at full stretch upon the deck. Howell brought up his scimitar so that the grip was near to his ear and looked about for his next target.

Further along the deck, Sir Horace was standing on his hatch cover, and aiding the defense as best he could. "There's one on the left. You two, get these here in the middle. Swing that club, I say, swing it. With the knives, stab, stab, stab!"

In the middle of the ship a pirate rolled over the side and onto the deck. The crewmen next to him recoiled in horror.

"Kill him!" Sir Horace yelled, pointing with his rapier. "Now!"

The pirate swung his sword in a wide arc around him. He did not hit anyone, but it was sufficient to drive back the defenders, and make a space. Other pirates quickly used the opening. They came, singly, in twos or threes, tumbling onto the deck.

It was all as Howell had feared. The sailors were just no match for the seasoned cut-throats pouring over the rail. One of Sir Horace's chosen men was on each side of this circle of frightened crewmen and tried to hold them together, but one of the sailors lost his courage and ran. Those around him fled as, one by one, the man beside him broke. The mob of terrified crewmen ran and stumbled and blocked one another's way, fleeing towards the seeming refuge at the aft of the ship.

The *crack! crack! crack!* of musket fire came from the pirate ship.

*Boom!* The smoke burst forth from the murder along with multiple cries of pain from the pirates still awaiting their turn to board. Ropes in the pirate ship's rigging snapped. Their sails were holed in innumerable places. Despite all the execution that the murderer had done, a dozen or more of the pirates were climbing up the side. None of them was grinning now.

The two men who had originally manned the little cannon now lay upon the deck. The First Mate groaned and barely moved, while the other was entirely motionless. The pirates still had their muskets, and they still wanted to shoot down those who operated the murderer, yet the Captain was bravely crouched behind the gun to aim it. Incredibly, beside him Howell's red-headed Rosy held the ammunition.

Sir Horace jumped off his hatch cover. He used his off hand to draw a dagger from a scabbard in the small of his back. It was shorter than his rapier, but just as thin and deadly. He began to back towards one of the sets of stairs that lead up to the quarter deck. He could not deflect the pirates' far heavier blades with his long, thin rapier, instead he ran through any of them who got close enough to try to strike him. Any that got closer than that, he dispatched with his dagger. Beside him was one of his men, who fought with an axe.

Howell was at the foot of the stairs opposite the one that Sir Horace was defending. Beside Howell was the other of Horace's men. He had found a scimitar himself, so that he and Howell were both armed as were their opponents.

*Boom!*

Directly between Howell and Sir Horace, a mob of terrified, unarmed crewmen was being forced by the pirates into a tighter and tighter herd against the bulkhead directly beneath the quarter deck. Their only hope for survival was that they would be captured. They would be alive, but they would be taken to the slave markets to enrich the pirates. Dying might be the better option.

Howell and Horace's guard, elbow to elbow, were defending their set of stairs by slashing with their swords as did their attackers. A pirate tried to chop down upon Howell with a mighty blow, but Howell swung his sword from his side to crash into the middle of the pirate's scimitar. The blade flew to the side, and cut in to another one of the pirates.

Then then man fighting at his side went down, and Howell was alone. He did not have the time to speculate on what might have caused the disappearance. Immediately another pirate swung his blade and Howell had to leap back to avoid the blow. Another attacker crowded in beside the first.

"Beware!" Howell could hear Rosy's voice coming from up on the quarter deck. She was pointing to a pirate who had been pinning the crew against the bulkhead, but now had turned, and was slipping in behind Howell.

He glanced over his shoulder at the pirate. How could he fight both ahead and behind? Then he had the briefest glimpse of Rosy pointing a pistol with a look of fierce determination upon her face.

*Bang!* the pistol fired.

Howell immediately felt a violent punch in his back. Blood flew by his ear. The pirate behind him sank onto the deck like a handful of loose sail cloth.

Two pirates came towards Howell. The one in the lead had his arm cocked for a savage chop. Howell took a step forward, and sliced down with his scimitar. He sliced into the man's hand, and slid along the bones of the man's forearm. The pirate yelled in pain, and dropped his weapon. The second pirate, however, was hurrying up, and was already practically upon him.

Howell prepared as best he could for the coming exchange. He was on his back foot now, and in no position to deliver anything like a serious blow. Suddenly the pirate tripped, and fell heavily. Just over this man's back, Howell could see one of the crew lying, bleeding, upon the ship's deck, his hand around the pirate's ankle. He had done what he could to help a shipmate despite his own, perhaps fatal, wounds. Howell cut down on the pirate, and made certain that he kept as much downward pressure as he could as he slid his blade back towards himself.

Suddenly, Horace's man was at his side once more. Howell and his comrade came up behind pirates who were turned to keep the crew against the after bulkhead. Howell simply chopped his weapon deeply into the back of the neck of the first man he encountered. The man beside him did the same. Howell could see Sir Horace on the other side of the roiling mob, defending his set of stairs, repeatedly stabbing pirates with his thin rapier or his shorter, but equally lethal dagger.

*Boom!*

The murder was keeping up a steady fire. The Captain was aiming, and Rosy had returned to the gun's muzzle to act as loader. She had'd had the time to fire the pistol into the pirate that had threatened Howell from behind, yet had not missed a beat loading the miniature cannon.

Both Sir Horace and Howell had but one man fighting beside him, but they were making an impression on the flanks of the pirate host. There were far more pirates opposed to Howell and his three mates, but, improbable as it was, it seemed that the defenders were having an effect. The raiders had not expected any resistance worthy of the name, and instead they found themselves contending with dangerous foes.

Now pirates pulled crewmen out of the terrified mob to use as human shields. They held their scimitars in their right hands and wrapped their left arms around the neck of some unfortunate sailor. In this manner they advanced, armed and protected.

One of the pirates was the man with a black face that Howell had glimpsed earlier. His fingers and his arms and everything that Howell could see was black. Howell had never before laid eyes on anyone like this man. He did not know what to make of him, nor did he have time to ponder the matter. Howell would hold his ground just he had when a boy from a rival gang had tried to steal Howell's loot. He would fight. He would win.

Three pirates with human shields advanced upon Howell. One after another they struck at him, and Howell was able to deflect each of their blades in turn. He used each of these blows as an opportunity to nick or slice the arm holding the sword before he had to shift his attention to the next attack.

*Boom!* The murderer fired down into the pirate ship yet again.

The black pirate did not have a human shield. He needed both his hands to hold a musket. He stepped up behind the three men that Howell held at bay. In an instant he would have the musket on target and fire. At point blank range there could be no possibility that he would miss. The muzzle was coming down.

"Sailors! Drop to your knees!"

These sailors spoke English, so Howell could speak in a language the pirates could not understand. Fortunately, the three human shields had kept their wits about them, understood what Howell wanted from them, and dropped to the deck.

Howell leaned forward and delivered a terrific blow to the forehead of the pirate that stood between him and the musketeer. The man collapsed. The musket continued to lower to take aim. Howell could look into the black void of the musket's barrel about to point directly at him. He chopped again, sliding his blade along the fore stock of the musket, mangling the man's first finger and slicing several more. The priming powder flared and smoked, but in his pain, the pirate dropped the musket and it fired into the deck between Howell's feet.

Howell took a step forward, slashing left and right to cut down the other two pirates. Then he finished off the black man who had held the musket with a chop deep into the man's chest.

Horace's man had been dealing with his own fight, but now he rejoined Howell. The sailors that Howell had freed picked up the scimitars left on the deck, and positioned themselves on either side of them. They let Howell and the guard take the lead, but they at least put up a good front. Slowly they moved forward. Howell and his mate struck whenever they could find an opening, and fended off blows as required. Howell could see the top of Sir Horace's hat as he too pressed in upon the pirates.

Howell noted that there were not as many of the bloodthirsty bastards as he had expected. There might yet be a chance to win this fight.

Pirates turned from guarding the sailors at the bulkhead to come in upon Howell from two sides at the same time. He chopped this way and that as fast as the heavy scimitar allowed. The sailors around him poked at their foes with the swords they had salvaged. Sailors from behind the pirates did what they could with belaying pins they had recovered or the knives some still held.

*Boom!* The Captain and Rosy did further damage to the pirate's vessel. Their sail was full of holes and many of the ropes of their rigging were dangling uselessly. The deck was a shambles. Hardly a man was on his feet in the pirate's ship.

Howell could make out Sir Horace on the other side of the deck, methodically impaling pirates with his blades. Whenever a pirate raised his sword to chop at him, he simply stepped forward, and slid his rapier between the man's ribs.

"Turinyow!" cried someone amongst the pirates. "Denan! Turinyow!" Howell had never heard this language before, but these words seemed to mean 'retreat', as there was a general falling back by the pirates to the ship's rail. Some of them tried to salvage what they could from their bloody defeat by bringing along their human shield to enslave, but Howell and Sir Horace moved to prevent that.

Across the width of the deck, Howell saw Sir Horace seeing off one of the pirates. The pirate's human shield was still in place, and Sir Horace heedlessly thrust his rapier through the sailor's arm to kill the pirate. Howell could not believe what he was seeing. Could this man be so cold-blooded that he would wound his comrade to achieve his goal, or was it just an accident?

The pirates began to scramble over the side and back down onto their ship. The vessel was beginning to drift away, what with the damage done to its rigging by the murderer and the wounds suffered by the crew. More than a few of the pirates fell into the sea in the gap between the two ships. These thrashed and yelled and drowned.

Sir Horace grinned and leaned over the rail. His face plainly revealed the satisfaction he felt.

It was over. Howell turned to look around the ship they had defended. The dead and the dying and the wounded lay everywhere. The smoke hung heavily from the repeated firings of the murderer and the pirates' muskets. On the quarter deck, the Captain grinned through a face begrimed with powder.

Redheaded Rosy beamed down at him, displaying the ramrod that she had used to load and reload the murderer, over and over. She could never imagined that she would be required to do something so distant from everything that she had experienced in life at the Countess's court, yet she had done it. She had seen what needed to be done, and had not hesitated to step forward and fill the dangerous post that their lives might depend upon.

Howell raised his arms, displayed his scimitar, and barked in triumph, *Oow! Oow! Oow!*

# Chapter II

## The Big Monument

During the night, the crew did what they could for their wounded shipmates, while the pirates went to their eternal rest amongst the sharks and the crabs. In the forenoon of the next day, the ship found its assigned dock in the bustling anchorage of Genoa. While the disposal of the pirates was done without art, making port demanded considerable skill to accomplish. The space between the piers was not generous, and there were many other vessels berthed on both sides of the narrow waterway. The sky above their heads was choked with masts and spars and gaffs and countless fathoms of rope. The stew-pot that was Genoa's harbor festered under the rising Sun.

Howell stood fascinated by what he saw, though he was sweating in the Italian Summer and his English wool. His breeches and common grey doublet had served very well during the windy voyage from England, but now he was sweltering.

He would rather have dressed like the listless dock workers that lounged on the pier beside their vessel, hoping for a day's work unloading the newcomer. Most of these wore nothing but knee-length breeches and perhaps some rag tied around his head. Their garb might be more comfortable than Howell's, but he knew their lives were not. He had lived like them. He had scratched for crusts, and slept in doorways until the day that Mr. Riss had taken him off the streets to do the necessary work at his counting house. At least in the premises of A. V. Riss and Company he was regularly fed and could sleep in the back room where he was out of the weather.

Eventually kindly Mr. Grace had taught him how to read and write, and cranky Mr. Riss had thought it necessary that he learn sums. He had had been lucky, but he had also had drive and abilities, which had given him wages and prospects. Now he was on his first voyage to supervise the cargo, and to deliver it to the agent that A. V. Riss and Company had in the city. He felt proud to be part of the administration of the ship. Howell was nearly a head taller than anyone else in the party, so he took a young man's pride in his physical gifts, as well. He would never again live like the paupers he saw on the dock.

He had just aided this Sir Horace fellow in a dangerous encounter as well. Who knew where that new connection might lead him?

Behind the pitiable stevedores, a considerable city was wrapped around the water front. It was not like any other town that Howell had ever seen. The white stone buildings with their red tile roofs crept up the sides of the sharp hills that crowded the water front, and extended back into a line of jagged mountains beyond. The docks and streets crawled with shuffling processions of faceless porters weighed down with crates, sacks, coils of rope, and the endless, numberless barrels.

Howell turned his head to see Rosy emerge from the after cabin. She was wearing a fresh wardrobe featuring a bonnet over her brilliant hair and a linen dress, so she looked every bit the lady that Howell had first seen when they had set sail. During the voyage they had spoken and learnt something of one another, but now they shared something far more gripping. They shared a very unusual respect for one another. She had seen the beast emerge from within him, and was not repulsed. Howell had seen astonishing courage in this woman, and was captivated.

She walked up to a place beside him, and spoke unnecessarily loudly. "May I see the books, Mr. Davidson?"

"Certainly, Miss." Howell held out the ledger filled with columns of words and columns of numbers. He could not imagine why she should be interested in all this.

She leaned close to Howell as though to get a better look at the information. She whispered to him, "Meet me at the big monument this evening."

Howell instantly realized that she was putting on this show so that no one would question what she was about. Howell pointed at a column in the ledger, and kept his voice low as well. "Big monument? What is that?"

"I do not know. There is always some big cross or statue or some such in the middle of town." She stood up straight, and smiled at Howell. Once again, her voice was overly loud. "Excellent. It appears that everything is in order."

"Miss?" It was one of Sir Horace's guards. "Perhaps you would like to go to your residence now?"

"Yes, that would be most satisfactory." She gave the man the full nose-in-the-air treatment, and did not even glance at Howell. She wanted everyone to think that Howell was beneath her attention.

Her escort extended a hand to help her off the deck and down the narrow gangplank.

Rosy hesitated. Was she really going to let this unwashed menial touch her?

Howell thought she was putting on quite a show.

Finally she realized there was no practical alternative, and took the man's hand. She went down to the pier, and then instantly withdrew her hand. She allowed Horace's man to guide her into the city, so long as he kept a respectful distance.

Horace had a word with his other guard, the man who had fought at Howell's side, and that man went down the gangplank as well.

Howell watched Rosy moving off to her new residence, and rolled her words around in his mind, tasting it. 'Meet me at the big monument this evening.' He could not be sure what was going on, but he would find out tonight.

Sir Horace walked up. This man and Howell had fought side by side, and beaten the pirates. They had worked together to save the ship. They were comrades in arms.

Sir Horace looked at Howell. "The papers if you please, Mr. *ah...*"

"Davidson, Sir." Howell was not happy. They had spent months at sea together, and the man had not learnt his name.

"Davidson, yes, of course." Sir Horace extended his hand with a slight smile on his face. "Mr. Davidson."

Howell handed over the folio smartly. He was skilled, he was clever, and, though he might yet be a young man, he imagined that he was already putting his mark on the world. He felt the pride rising within him as he began to think of how he had acquitted himself in the recent battle. He also thought of the odd scene of Sir Horace stabbing one of their own crewmen to get at a foe.

Sir Horace handed the folio that contained the ship's papers back to Howell. "I presume that you will be supervising the unloading of the ship?"

"Yes, Sir. The first of the cargo is about to come out of the hold now, Sir."

"Very good. Get on with it." Sir Horace turned, and walked leisurely towards the quarter deck.

Part of Howell's job was to put a good face on the services provided by Mr. Riss and Mr. Grace, and he thought he had done very well with the obnoxious Sir Horace. Howell moved to the center of the ship and 'got on with it'. It was Howell's responsibility to account for each item that came out of the hold.

The crew had the hatch covers off, and prepared to hoist out the first of the cargo. It was only then then that Howell noticed that things on the pier were being supervised by Sir Horace's man. He had collected a crew of stevedores on the dock, and was keeping everyone else away. He had assembled quite a number of wagons, as well.

A rope went down, and was attached to the first crate in the hold. Three sailors hauled on the rope, putting all their weight onto it, but nothing happened. Immediately two more crewmen were called to help. Whatever they were drawing out of the hold must be unusually heavy. When the crew finally got the load high enough that Howell could see it, he found that it was not the barrel one might have expected, but rather a wooden box about the size of a coffin. It had a peculiar symbol burnt into it; a rectangle with a line through the middle running the long way.

One by one, these crates emerged. Each had the mysterious symbol burnt into it. The crates were swung out beyond the ship's side, over the dock, and placed into one of the wagons. Howell noticed that only two of these crates was put into each wagon, which left a lot of room above them. They must be heavy indeed to use so little space. They were placed directly into the center of the wagon box so that the weight would be spread evenly between the front and the back wheels. When a wagon was filled in this fashion, it was dispatched. Then another wagon was moved into place, and the work began all over again.

Sir Horace watched the entirety of this process from the quarter deck. He must have been doing a mental count, since he turned and walked away as the last of his crates was pulled out of the ship. He displayed no emotion throughout the period required to complete the unloading.

Sir Horace's wagons rolled away, and Howell watched as the rest of the ship's contents came out, barrel by barrel, and crate by crate. At last it was done. He consulted the papers in his hand and confirmed that everything had been dealt with. He was free to pursue other business. He was going to pursue Rosy.

He walked down to the dock, and stepped out into the street. The Sun was sinking. Where to begin? He had seen nothing from the ship nor the docks that hinted at something that might be referred to as 'the Big Monument'. Rosy had said that such a monument was generally the focal point of the city, so Howell turned his steps towards the middle of town. Had she taken into account the dangers she might face in an unfamiliar city? He had seen her bravery, but could she also play the fool?

He was initially on a rather small street, but he quickly crossed one of the innumerable bridges over the innumerable canals, and came to one of the city's larger streets. He began to follow it. He found that Genoa was much like London in this regard. Leading off from the main thoroughfare were twisting alleys that Howell found all too familiar. There were churches, towering and grand.

There were proud buildings that Howell assumed were the residences of the famous merchants of the city. Might any of these be the place where Howell was supposed to meet Rosy?

Wagons passed him loaded with everything from the furniture of one of these grandees of business all the way down to horse shitte. There were beggars and peddlers. There were women that you could have for the right price. Shops were everywhere, trying to draw customers with signs painted with some manner of symbol that advertised their wares to those who could not read. Howell hurried passed them all, concerned that Rosy's plan might not have been as well thought out as she had imagined.

Then he saw it. He stepped out of the great avenue he had been following, and was confronted with a spectacular fountain. Dead center of this was a huge cross. Surrounding it were twenty, perhaps as many as thirty streams of water that emerged from the perimeter of the fountain and arched inward towards the Crucifix. The breeze caused them to leave some mist on the downwind side. They made him think of the petals of a flower made of water. Impressive structures made entirely of white stone walled off the square surrounding the fountain. Off to one side there was a grand statue of a horseman. There could be no doubt that this was Rosy's 'Big Monument'.

He found a seat near the fountain that gave him a good view of the statue and all the rest of the square. A few peddlers had carts here and there. Strollers enjoying the cool of the evening walked past him. The shadows crept across the square. The fountain's water splashed. A small pack of ragged boys stepped out of an alley, and cast about for a likely target for their thieving. Their leader pointed, and they started off.

Howell began to think about what strange turns life could take. When he had lived like the boys that he had just seen, who would have believed that he had be transformed into an educated man? The skills that he had learnt in the alley with a bit of broken glass wrapped in a rag had transformed effortlessly into skill with a scimitar. Now as a junior clerk, not even worth learning his name by some people, he was asked to rendezvous with this stunning beauty. What was next? What was hidden in this broad world before him that would yield a surprise or an opportunity?

The Sun moved ever closer to the horizon. Howell laughed to himself. Was not just like a woman to be late for her own appointment? Perhaps it was worse than that. Perhaps she had only been toying with him, and had never intended to come to 'the Big Monument' at all. Perhaps it was much worse than that. Perhaps Rosy had dreadfully miscalculated. Perhaps she was lying robbed and bleeding in some filthy, desolate street in the depths of the city.

No! There! A fetching figure approached him from the far end of the square still clad in the linen he had seen her wearing on the ship. She was accompanied by a man that Howell had never seen before. Rosy's hips swung irresistibly.

Howell grinned. He had found the right place and she had come to him without mishap. She walked up to him languidly. He gloried in her hair, her eyes, her bosom, and the curve of her hips. Howell rose to greet this delectable creature and reached for his hat. He swept it off in a graceful bow as he had seen gentleman do.

"Good evening, Mr. Davidson. What a pleasant surprise to encounter you."

Surprise? Howell was at a loss. This was her idea in the first place, and now she was acting as though their meeting was completely a matter of chance. What was going on here?

She turned to the man who accompanied her. "This is Mr. Howell Davidson, a gentleman I met on my recent voyage. I thank you for your service, but I am sure that Mr. Davidson can escort me now." She slipped her hand into Howell's arm. "This is the butler from the residence that has been provided for me and my party. He was kind enough to escort me while I took the air, but now you can continue."

"You honor me, Miss Kincade." Howell instantly fell in with her plot.

The butler displayed a small bag. Doubtless, he had been carrying it all this time, but Howell's attention had been otherwise occupied. "Would you like this, ma'am?" The butler offered the bag.

"Oh, yes," Rosy said. "I'd completely forgotten about that. Mr. Davidson?" She nodded to indicate what she wanted from Howell.

Howell took the bag from the butler. It was not even as big as his fist. Then he extended his hand again. "Thank you, Sir."

The butler had never expected to be treated so well. He shook Howell's hand tentatively.

"Good evening," Rosy said to the butler.

"Good evening, Miss." The butler gave a little bow, and started off across the square.

"Let us be off then," said Howell. He turned their backs upon the butler, and began to walk. "And where is it that we are going?"

Rosy looked up at him with the hint of a smile. "I believe that we should walk beyond the Palace of the Doge." Rosy pointed at a huge, white building.

"The who?"

"The Duke. The ruler of all that you see about you."

Howell looked away and muttered, "How nice for him."

It was getting dark enough now that braziers were being lit to provide some light as the Sun made its slow descent. Rosy continued to lead Howell along the dark streets ever deeper into the city.

At last, Rosy broke the silence. "The butler you just met is a perfect example of the stiffness of everyone in the household. 'Must do this.' 'Cannot do that.' It is maddening." She turned to Howell to see how he was taking this. "You cannot imagine how suffocating it feels. I seize every opportunity to do something, anything, to feel freedom."

"So I am your current amusement, then."

"Oh, my goodness!" She used her free hand to cover her mouth. "That is not what I meant at all!"

"And what was it that you did mean?" Howell was calm, but looked at her in a way that demanded an answer.

"Well..." *emm* She looked away at the street they were walking. "Mr. Davidson," she still was not looking at him. "I saw you battling those pirates."

"Yes?"

She gathered herself, and looked again at him. "Your actions spoke to me. I saw your courage. I saw your presence of mind to trap the pirates by advancing on one side of them at same time that Sir Horace was moving in on the other. I saw someone that I wanted to know." She looked at him with stunning intensity. "You promise to be someone unique."

He could not think of anything to say, nor what the word 'unique' meant. The way she said it though seemed very encouraging.

They emerged into another square. This one had no fountain, but it had some braziers and benches all around the periphery. Howell led them to one of these latter, and they sat rather closer than polite society would approve.

"I wanted to talk to you about something else, as well." She put her hand on Howell's.

Howell could think of any number of topics that he had like to probe deeply with her, but that would have to wait.

"Yesterday, I helped the Captain fire that little cannon over and over. He…" she bit her lip "…I should say we, must have killed and wounded many men on that ship."

"Yes you did."

She hurried on. "That was all at a distance, though. I could not see them. There was all that smoke, and I was busy with the powder and the bags of shot." She paused, looking at him. "There was one man, though, that I saw was right behind you. I had to do something. The Captain handed me his gun..."

"Pistol." Howell immediately regretted correcting her, but she did not seem to take any notice of it.

"Yes, pistol. He said it was ready. The pirate was right in front of me. I could see the rag tied around his head. I pointed the gun directly between the man's shoulders, and pulled the trigger. There was a terrific *bang*, and a big cloud of smoke, and the gun flew back out of my hand." She paused, not sure of what Howell thought of all this. "I must have shot him. I may even have killed him."

"You did shoot him. You shot me as well."

She did not even gasp, but just sat there, staring at Howell, with her mouth hanging open and eyes as big as saucers.

"The ball went straight through that bastard, and hit me in the back. I have got a bruise, but I am happy for it. Your shot almost certainly saved my arse."

She gathered herself, and thought for a long moment. "I do not know how I should feel about what I did."

Howell did not either. For him, the thrill of confronting an enemy, and tasting victory was just a given. He could imagine that this might not be the case for a lady, though. "I suppose it depends upon the person. It was certainly something well out of the ordinary for you, but I cannot see it from your angle. Some might think it bad. Some might think that doing whatever you could to help your mates was exactly the right thing to do."

She smiled back, comforted by what he had said, relishing the moment. She even moved her hand as though hefting the pistol again. Then her facial expression changed to the sinful. "I have it in my mind to do something else well out of the ordinary for me."

"Really?" Howell's base instincts made themselves felt.

"What do you think I have had you carrying around?"

Howell looked down at the bag in his hand. He had not given it any thought since the butler had given it to him. "I have not the first clue."

Rosy had a teasing sort of grin. "It is a treat for us."

"What sort of treat?"

"Go ahead, open it up."

Howell untied the bow holding the bag closed, and peered in. The smell was new to him, but not disagreeable.

"What do you think it is?"

"A magic potion." He said it as though he knew what he was talking about.

"No, you devil!" Then Rosy reconsidered. "Well, in a manner of speaking, I suppose you could think of it in that way."

"Can it make you kiss me?" Howell had never found restraint to be a virtue.

"No." She did not pull away, however, and she kept her eyes fixed on his. "But it can make money appear out of a puff of smoke."

Howell shifted his attention to the bag. "That sounds pretty magical to me. What is it?"

"It is some leaves from a plant they have been growing in Virginia."

"In America?" The English colony at Jamestown was barely a few decades old. It was a tiny foothold in a seemingly endless New World. He was holding something that had come from a land that no one had even suspected existed not so long ago.

"It is called tobacco. It is all the rage in England, and the people who sell it are making handsome profits."

Howell gingerly opened the bag, and inhaled deeply of the unique aroma. He wet his finger, and moved to stick it in the bag for a taste.

"You do not eat it!" She slapped down his hand.

"What? You sprinkle it on your head like fairy dust?" He wiggled his fingers over his head.

"You've missed by a mile." She laughed as she produced an odd clay tube. It was very long and narrow for most of its length, but at one end it turned an almost square corner, and flared out to be big enough to put one's finger in. "This is called a pipe. You put the tobacco in here…" she said indicating the big end, "…and set fire to it."

"What? You burn it like incense?" He had seen incense burned in church parades, but that was in large brass pots. This clay thing, this pipe, was tiny by comparison.

"Something like that." She took the bag from Howell and suddenly looked over at him. "How are we going to light our little fire?"

Howell gazed into her eyes and then glanced around the square. "I will get something from one of the braziers over there."

"Would you take care of that, dear?" Rosy set to work getting the tobacco into the bowl of the pipe.

Rosy's tone had been pleasant enough, but being given instructions like this brought up unpleasant memories of being told to fetch as he had been in his early days at the counting house of Riss and Grace. He had brought in packages, reams of paper, and firewood. He had hated it all. What he was doing now was something different. He was not being ordered so much as working as a part of a team. He liked it. He returned to Rosy with a stick with a bright ember on the end, and Rosy presented him with the pipe. He was at a loss. What did one do with this thing that he was holding?

It turned out that Rosy was just re-arranging her dress. As soon as she was comfortable, she took the pipe from him and held the bowl in the tips of her fingers. Rosy lightly placed the long, thin end to her lips.

Howell stared in bewilderment, holding the smoldering stick like a choirboy holding a candle.

"Light it."

Howell regained his senses, blew up the coal on the end of his stick, and placed the glowing end in the bowl. Rosy sucked mightily, the tobacco caught fire, and… *Hack! Choke!* Rosy began to cough violently. In her gyrations she let loose of the pipe. For one awful moment Howell could see the piece of fragile pottery suspended in mid-air, falling to certain destruction upon the flagstones all around them. Instinctively, he thrust out his hand and caught it. *Whew!* Then the heat of the burning tobacco made an impression on his hand, and he jumped. For another awful moment, it seemed that Howell would finish the job instead of salvaging the situation, but he managed to get a hold of the thin, cool end, and restore calm.

The two of them sat beside one another on the bench, their shared moment with them as well. Suddenly they were laughing with one another, with Rosy occasionally coughing.

Howell smiled at her. "That was a near run thing."

Rosy giggled a little, and wiped her eyes. She gave another little cough. "Well, it is not overwhelming, but I am feeling a pleasant, well, buzzing right now." She giggled. "It is rather nice."

Howell put the pipe to his lips tentatively. He drew on it carefully, and saw the tobacco flare slightly. He could only hold the smoke in his lungs for an instant before he had to expel it, but he managed with just one loud hack instead of the coughing jag that Rosy had experienced. She was right. There was something new and pleasing about what he felt.

Rosy took the pipe and tried her luck again, this time a bit less recklessly. She managed to hold down a considerable amount.

"You said that people in England are making quite handsome profits selling this tobacco?"

Rosy nodded vigorously while holding her breath. She managed a few heart beats longer than had Howell. Finally, she emptied her lungs with a great sigh. "Yes…" she grinned. "…they sell it to the sophisticated and to those who wish they were." *Cough*

*Ahh* Howell relished a prospect that he saw opening before him. "An interesting situation, Rosy."

"Rosy?" She was taken aback. "Are you referring to me?"

"Yes. That's the name I have given you in my mind."

Rosy smiled. "I rather like it, Howell. Just something between you and me."

Perhaps it was the tobacco. Perhaps it was the thought of future profit. Certainly it was her stimulating proximity to him. Sir Horace had been unable to recall his name. Now this stunning woman had accepted his name for her, and had used his Christian name without hesitation. Whatever it was that motivated her, Howell could see that their future together was growing apace.

She handed the pipe to him. Absentmindedly, he took it, and had another small pull. He gazed at Rosy with a joy that he had never experienced before. He had known women of course, women in the raucous brothels of east London, women who held their noses so high it was a wonder that they could get through a doorway. Something was different here though, very different. Rosy was friendly without being vulgar, she was dignified without being snobbish. She was brave without being undisciplined. He was fascinated by it all.

"It is been a lovely evening, Mr. Davidson."

"Howell."

"Yes, Howell." She enjoyed the intimacy of saying his name. "It is been lovely, but I really need to return to my lodgings."

"Suppose you sold some of your tobacco? Make some of that handsome profit for yourself?"

She looked around the square where they sat. "There is no one here to sell to, Howell."

"Not here. At the soldiers' camp. There are plenty of customers there."

A regiment of mercenaries that was marching from one army to another had broken its journey at Genoa, and was now encamped on the outskirts of the city. The soldiers had been paid, and they were everywhere, carousing and squandering their money.

Rosy looked at him for a long moment. "There will be low women there."

"Many."

"And coarse soldiers."

"Who will be drunk for the most part."

"Do you really think I want to go to such a place?"

"I think that you would be pleased to put silver into your purse." He paused while he let her consider this.

Rosy hesitated as well. She looked away from Howell, and bit her lip.

"You and I have stood together before pirates." Howell felt as serious about this as he sounded. "We can certainly stand together before a mob of drunks, and you may be enriched by them."

Rosy finally put her eyes upon him. She gazed at him briefly and stood. "Let us be off." She was aimed towards the city's gate. They walked, arm in arm, to another opportunity for Rosy to escape the stifling atmosphere of the court.

Howell's curiosity got the better of him. "I have to ask, how is it that your household will not notice your absence tonight?"

"It is not that they won't notice, it is just that they can do nothing about it."

Howell looked at her questioningly.

"I have been in the Countess Elizabeth's household since she was a little girl, in England. I was brought in as a sort of playmate for her, and we grew up together in King James' court."

"Our King James? Of England?"

"Yes, Howell, our King James. He's Elizabeth's father, but he's not just King of England. He began as King of Scotland as well, so he brought some Scots like me with him."

Howell whistled in appreciation. "I had no idea that I was in the presence of royalty."

Rosy jabbed him hard in the ribs, making Howell grunt. "I am not royalty, you twit, but the Countess and I are more than mistress and servant. That's why I am going over the Alps to join with her. When I shared with my faithful servants that I was going out tonight, they had to accept the fact."

"Excellent! Then let us make this night a truly magical one." Howell and his improbable companion issued from one of Genoa's gates, and made their way towards the wild, fire-lit encampment on the edge of town.

Howell was curious about something else as well. "How does one acquire tobacco, anyway?"

Rosy did not answer immediately. She glanced at Howell, and pondered her response. "You steal it."

"The Hell you say!" Howell had not been prepared for such a bold answer. He kept finding new things about this woman that he liked.

Rosy started at Howell's crude response, but proceeded with her explanation all the same. "There is a gentleman, and I use the word reluctantly, in the Countess's household who has rather more tobacco, and pride, than is suitable."

"So you decided to relieve him of some of his burdens."

"She smiled wickedly. "He is a mutual acquaintance of ours, you know."

"Acquaintance?"

"I have dealt with his self-important antics for quite some time, but you have only seen him fight pirates and count crates."

"Sir Horace." This man wanted to be obeyed and Howell craved respect. Were these the same thing? Were these two sides if the same coin? Was this something he shared with Sir Horace?

"In any event," Rosy continued slyly, "you have met."

"Did not he miss what you took?"

"Oh, he missed it right enough. He was fairly seething when he discovered his loss, the pompous ass, but there was little he could do."

"He could turn the household upside down!" Howell had seen Mr. Riss cause an uproar in the counting house over the loss of a single quill. "He could line everyone up, he could…"

"No, no…" cooed Rosy, "…no he could not. The Countess's father, King James, just hates tobacco, and Elizabeth tries to put up a show of following his lead. She really doesn't care one way or t'other about Sir Horace's smoking, but she would never put up with some kind of inquisition for lost tobacco in any household under her authority. With all the puffery and preening and secret news whispered in dark corners that occupies his days, Sir Horace dropped his guard, and I was able to take advantage." The triumph positively glowed on her face.

They emerged from vibrant but conventional Genoa and entered a new, spur-of-the-moment world. Where a week ago had been an empty field, tonight was a town of tents and wagons.

Kiosks and stalls of every shape and description surrounded them. There were come-ons from hawkers and the haggling of customers. The company consisted of men from many lands, so there was wine and beer to drink, bread for sustenance, and cheese to flavor it all.

"Where will you set up shop?"

It was Rosy's turn to be caught off guard. "Set up shop?"

"If you are going to pull scores of grosso out of that tobacco pouch of yours, you must make a stand somewhere. No one will buy unless you sell."

Rosy came to a standstill. She stopped luxuriating in the atmosphere, and looked around the swirling fair. Nothing seemed to strike her as a likely venue.

"If I might suggest, Your Highness?" Howell swept off his hat in an operatic bow.

"Yes, my Lord?" she replied with an elaborate curtsy.

"If you will allow me to guide you, I think that I will locate just the right location to fulfill your requirements." Howell led off into a portion of the field that they had not yet explored. He was looking for a place with a small crowd and a proprietor who would welcome something to draw customers. Wealthy customers with plenty to spend for preference.

They came upon a wagon where wine was being sold, and a small group of well-dressed customers chatting happily amongst themselves. Howell walked up to the proprietor. "Good evening, Sir. My name is Howell Davidson, and I should like to discuss a business matter with you."

"Why would I give a shitte?" The wine seller did not even bother to rise from the keg he sat upon. He had an unmistakable French accent.

Howell did not hesitate to respond, "Because you are doing such a poor trade, monsieur." Howell took a puff on the pipe.

The wine seller bristled at Howell's comment. "And what the Hell is it to you?"

"Nothing, Sir, nothing at all, but I thought that if I provided something new and enticing, sold by a beautiful woman, perhaps it would draw additional customers, and make a difference in your cash box." Howell looked around at the crowd passing by. "Of which you are in desperate need."

By this time, the men who were drinking this man's wine were watching the conversation with considerable interest. Some were also taking an interest in the smoke that Howell blew amongst them, and they were all were taking an interest in Rosy.

"And who are you to tell me what is happening with my trade?" The Frenchman got angrily to his feet, but then he paused and chewed his lip as though considering what Howell was offering. "What is that shitte?"

"My I ask who I am addressing, Sir?"

"My name is Ruzeque, and I asked you a question."

"Yes, Monsieur, you did, but I think that it would best if Miss Rosy answered it."

Distracted by the banter between Howell and this Ruzeque fellow, she almost missed her cue, but seeing all the men looking at her, Rosy quickly recovered. She held up her pouch. "This is the greatest sensation in the world."

Ruzeque frowned, and cast a disapproving look at Howell.

Howell smiled back.

Rosy's rather weak beginning did not stop her. She pressed forward with increasing confidence. "It is the latest discovery from America! It pleases the senses, aids the digestion, and calms the nerves."

Howell nudged Ruzeque. "Pretty good patter, eh?"

"Nothing wrong with her looks, either."

Hearing this comment, the color of Rosy's face matched her name, but she soldiered on. "It is the newest thing from the Indies! This is tobacco!"

"Tobacco?" Ruzeque began to show some real interest. "I have heard something about this crap. You breathe the smoke, am I right?"

"Indeed, Sir, you do." Howell tapped out the burnt residue from the pipe and handed it to Rosy.

She displayed the pipe. She turned it around so that everyone could see what she was holding. She took a pinch of the tobacco from the pouch, and began to pack it into the bowl. "One places the herb into the pipe in this fashion."

Ruzeque scanned the crowd to see how his customers were taking this.

"Then you start it burning."

Howell had taken a brand from the fire in front of Ruzeque's wagon while Rosy was talking, and now stepped up to her side smartly. He presented it to her. Rosy took it theatrically, placed the match in the bowl, and inhaled on the pipe. *Cough! Hack!*

"Oh, this is going to be a terrific crowd-pleaser." Ruzeque turned to Howell. "Come see people choking themselves! Free admission!" He stomped back behind his wagon's tailgate. "If you could just get a guy who hits himself on the head with a club, we could really pack them in!"

Rosy was humiliated by this Frenchman's slights, and furious with Howell for having brought her here to be the butt of such jokes.

Howell reached for the pipe. "Excuse me."

Rosy handed it over doubtfully.

"I will have a word with Monsieur Ruzeque. Everything will be fine." Before Howell could take a step, however, two of Ruzeque's customers blocked his path.

"I beg your pardon, Sir, but I could not help but overhear the lady there." The young gentleman looked sideways at Rosy, and she returned a very convincing half demure, half come-hither smile. "She said that smoke stuff is good for the digestion. And it is from the Indies." He said the last as though the words were a magic spell. His companion nodded enthusiastically.

Howell casually took a puff on the pipe, and exhaled the exotic aroma in the direction of the two men. "Would you like to try some? Two grosso each." Howell would charge more later, but he needed immediate success now.

"Two?" The men quickly consulted with one another. "How about three for both of us?"

"Done!" Howell offered his hand. They shook, and Rosy appeared at Howell's elbow, her hand discretely extended. "Oh, yes, Miss." The gentleman consulted his purse while trying to look her over, and yet not offer offense. "Three grosso. Here you are."

Rosy slipped off in the direction of Monsieur Ruzeque, displaying the coins to him, and sticking out her tongue. Howell offered the pipe to the men in turn. One coughed spectacularly, but the other held it down as if he had done it all his life.

There were several other people who had stopped to see what was going on, so Howell put the pipe in his mouth, and exhaled a cloud in their direction. He was getting quite a buzz now. "Ladies and gentleman, just arrived from America, I offer you tobacco. The perfect accompaniment to a glass of fine wine." He swept his arm towards Ruzeque's establishment.

"That's more damn like it."

One person in the crowd stepped forward, and the night's real business began. Between the unusual aroma, Howell's constant patter, Rosy's subtle attractiveness, and a crowd drawing more of a crowd, the action was non-stop. Gentlemen, soldiers, and women were all curious. Enough of the tobacco customers asked for wine that Ruzeque actually ceased complaining for passably long stretches.

"It looks like you are really raking it in, Davidson."

"Well, so far, there has been no discussion of my wages." Howell gave a questioning look at Rosy.

She was in the midst of counting some coins, caught off guard, and embarrassed. Behind her freckles, her face turned the color of her hair. The truth of his statement could not be doubted. Howell was doing all of the work, and yet she had all the money. She recovered quickly, though. She placed her hand lightly on Howell's arm, and gazed into his eyes. "Howell, dear, fear not for thy contentment."

"My God." Ruzeque rolled his eyes in an even more exaggerated fashion than he had the first time.

As it had so many times before, Howell's hastiness got the better of him. "Perhaps you might grant me a token of thy appreciation now. Nothing overly dear. Something I am sure you can spare."

Rosy looked at him warily.

"Kiss me."

She stood, shocked at his boldness. As she thought about it though, she realized that this was the sort of uninhibited action that she had fled when she had left her current residence. She stepped forward, put a hand on either side of his face, and pulled him towards her for a brief, but pleasing, kiss.

"Hey! Turtle bloody doves!" Monsieur Ruzeque pointed at a small group that had just wandered up. "You got customers!"

Howell and Rosy returned to their routine. Rosy's tobacco pouch was emptying fast, and the other pouch that hung around her neck gained the weight of many coins.

Then the inevitable happened. One of their customers coughed so violently that he knocked the pipe loose. It went cart-wheeling into the air, and slammed right into one of the spokes on Ruzeque's wagon. The shattered bowl fell to the ground, but the stem simply disappeared. Howell and Rosy looked down at the end of their business venture.

Then they looked at one another. It had been wonderful. It had been the first thing that they had ever done together. Something shared. Something grand.

Someone in the crowd called out, "Hey, I want my turn!"

Howell looked over Rosy's head towards their audience. "There's no more tobacco, folks, but there's still plenty of good wine to be had." The crowd began to disperse as only a few of them had decided to follow Howell's suggestion, and patronize Ruzeque's wagon.

A few tears began to slide down Rosy's cheeks, and then she hugged Howell gently.

Howell was more than happy to return the gesture.

She sniffled. "There is nothing to be done now, dear." She wiped the tears off her cheek with the back of her hand. "All we can do now is move on."

"And drink," said Ruzeque, handing a mug to Howell. "And I do not want to hear any boo-hoo-ey from your stink-weed lioness after she has made a shitte load more money than I did."

Rosy disengaged from Howell, and turned to Ruzeque. "I am sorry monsieur. I did not realize that you were not doing well." There was real sympathy in her voice.

"Save it, cheri." Ruzeque shoved one of the mugs in front of her. "I have not done such a good business since that time two bawds got into a wrestling match in a mud puddle in front of my wagon." He raised his mug and looked at Howell. "Here's to the shared experiences that pull our jolly ragbag of thugs and perverts into a family."

Howell raised his in salute. They both looked at Rosy. She swept back her hair and seized the heavy mug. She raised it in salute to the two men.

"You two lambs did a first-rate job, damn your eyes." Ruzeque drank deeply.

Rosy and Howell let the warm, fruity liquid slide down, and spread its relaxation through their tense limbs. They smiled at one another, and began to think of what might come next. Rosy sidled closer to Howell.

"You are not going to commence snogging again, are you?" Ruzeque smacked his lips in an outrageous parody of kissing. "I am drinking on an empty stomach you know."

"That remains to be seen." Howell winked at Ruzeque. With all the tobacco and now wine, too, he was feeling no pain. He put his arm around Rosy's waist, and pulled her close. "So how much tobacco do you have left, dear?"

Rosy set down the heavy mug. Her face beamed in contentment. The evening had supplied its share of success, after all. "Well, not a lot really. Perhaps a third, maybe less."

Ruzeque slammed his mug down onto the wagon's tailgate. "What? What? Are you saying that we had to suffer your caterwauling when you'd already collected more money than Midas and still have product left?"

Rosy looked at him sheepishly. "It is just that without the pipe…"

"Without your damned pipe you'll be forced to stop working, drink free wine, and enjoy yourself! God's nails! You are tearing my bloody heart out."

Despite herself, Rosy smiled. Then she looked at Howell, beaming beside her, and suddenly the two of them burst into laughter as one.

"Monsieur Ruzeque is right," Rosy looked up at Howell, "but enjoyable as this is, I did not want it to end."

Howell collected her into his arms. "Oh, never fear, it has not ended yet."

Rosy looked up at him in confusion.

Howell shifted his attention to Ruzeque. "You sleep in the back of the wagon, am I correct?"

Ruzeque eyed him warily.

"Yes or no?"

Ruzeque thought about his response for a moment. "Oui."

Paying no further attention to Ruzeque, Howell motioned towards the rear of the wagon. "Rosy. Let us count what you have earned." He placed Ruzeque's keg to use as a step so that she could get into the wagon.

She hesitated. What were his intentions? How far out of her normal life was she willing to go? Nicotine, alcohol, silver, and passion brewed a potent elixir. She pulled up her skirt, and put her foot on the keg.

Howell helped her get onto the wagon. Then he climbed in after her.

She turned and knelt before him in the wagon box. They both had to kneel to fit into the tight space, but this served only to enhance the atmosphere. He looked at her red hair, glowing in the feeble light that came in through the canvas. He looked at her eyes. She was watching him. She was wondering what was going to happen next. He leaned forward, gently put his hand behind her neck, and drew her to him. He rested his forehead on hers.

She looked up, and put her lips on Howell's. "When you do business, you get paid," she murmured.

They kissed again. And again.

There was something here that was new to him, a feeling that this kiss was not just about their bodies, but something more. Something that Howell imagined could be described as one's soul.

Rosy pulled back, still looking at him with an intensity that Howell felt meant far more than just a few kisses could satisfy. She took the leather thong from around her neck, and drew out her purse from its resting place between her breasts. "Monsieur Ruzeque was right about sharing out the money."

She upended the purse, and dumped the mass of coins onto her skirt stretched between her thighs. A delicious clattering sound filled the space between them.

"Damnation!" Ruzeque exclaimed from outside the wagon.

She tossed a coin to Howell and dropped one beside herself on the blanket that Ruzeque used for his bed. "One for you, and one for me." She tossed another to him, and another, and another.

Today we will share money, thought Howell, but presently we will share something more, something much more.

# Chapter III

## Revelations

Howell and Rosy walked arm in arm through the maze that was Genoa's streets. If anything, the night was even darker than it had been when he had led her to the soldiers' camp, and they had gone into the tobacco selling trade. The night was as cool as it ever was in this city. They did not mind that they only made the slowest progress, rather they were content to remain in one another's company for as long as possible. They were silent. Each of them was considering what had just happened that evening.

"Well, you were looking for something out of the ordinary, were you not?"

Rosy did not echo Howell's light tone. "It was more than that, Howell." She did not look at him, but clutched his arm tightly. "I am not sure what happened, but it was not just some excursion from my normal life."

Howell was not sure either. Whatever it was that he felt for this woman was something out of the ordinary for him, as well.

They arrived in front of quite a grand building. Would he ever be let into somewhere like this? Rosy took one step onto the stairs, and stopped. Howell gathered her in. They kissed. Rosy was less tentative than she had been just a few hours before.

The door swung open, and there stood the butler. He was relieved beyond words to see Rosy returned safely.

"I have to go in now." The words came out, but she did not move.

"And I have a cargo to account for."

They gazed at one another for a long moment.

Howell and Rosy exchanged one more happy look, and Rosy turned away. The door closed behind her.

He collected himself. He changed his focus. He had work to do. He aimed downhill towards the waterfront. His first order of business would be to find a man named Storey, the agent that Riss and Grace maintained in Genoa. However, it would be many hours before the Sun came up, so this Mr. Storey was doubtless still snug in his bed. What should Howell do to pass the rest of the night?

He really did not have any choice but to go to a tavern. No other place of business would welcome him at this hour. He passed a darkened shop with the three balls of a pawn broker displayed on the outside. Then he found a place with a peeling, white sign above the door. There was a mug painted upon it in black, and some red to indicate the wine it held. Howell went in.

No one seemed to be awake. There was a small fire in the grate to keep the foul pottage warm. Howell wondered what vegetables and scraps of meat came to hand in this city to be thrown into the pot.

He walked around the counter, took a mug, and turned the tap to fill it. On his way back to the far side of the bar, he gently kicked the stool that the publican sat upon.

*Hunh?* "What?" The man came to his senses, more or less. "What are you doing?"

Howell arrived at the front of the bar. "Doing some business."

"Business? Thieving, you mean."

"Not if I pay for it." He tossed a grosso onto the counter.

*Hmm* The host seemed in a better mood now.

There were two men with their heads on their crossed arms at one table, and at another was a man and a woman. All of them were snoring noisily. Howell found an unoccupied table, and sat down.

Several thoughts had been bouncing around in the back of his mind, and now he took the time to properly consider them.

As the battle with the pirates was beginning, he had heard Sir Horace say that men were going to die anyway, but if we let the ship be taken, 'it would all be for naught'. All be for naught. What did he mean by this odd phrase? He did not say that we will all die, or that we would all be sold as slaves in Algiers. Howell tried to think through all the possibilities. Perhaps it was a completely innocent choice of words. All would have been lost, our cargo, our freedom, perhaps even our lives. Perhaps.

None of this rang true with what he had learned about Sir Horace. It seemed much more likely that he meant that if they could not keep the ship, something that he had worked on very hard would go into the cesspit. Maybe it was not the ship at all. Maybe it only his part of the cargo.

Another picture of Sir Horace was lodged in his mind as well. Towards the end of the fight, as the pirates were retreating, he had seen Sir Horace drive his rapier through the arm of one of our own sailors in order to stab a pirate. It might not even have been the only time. Perhaps he had done something similar more than once.

Howell could not put his finger on it just yet, but it seemed to him that Sir Horace had something on his mind other than what Howell could see on the outside. Sir Horace was a driven man. There was little that he would let deter him from his goal. Goal. That word was something that Howell needed to firmly attach to his picture of Sir Horace.

Finally, the Sun was sufficiently high above the horizon that Howell thought that Mr. Storey must be awake. He left behind fully half of the awful wine that he had tried to drink, and walked along the jetty that encircled the harbor. On one side were the piers that jutted out into the water. Ship after ship was moored to these. Some were quiet, some were loading or unloading, and some just seemed to be engulfed in general chaos.

On the other side of Howell's path, street after street led off from the jetty and into the city. He had been told that the office he sought was in an inn called 'The Bull'. He arrived at the ship that had carried him from London, but he had not seen anything that hinted at the inn. It had to be up one of the streets that issued from the jetty, but which one? There was no guarantee that it was nearby. Genoa was a big harbor after all. Mr. Grace must have thought that 'The Bull' would be easy to find, so Howell was somewhat surprised that it was not obvious. Where could he turn for help?

Then he was approached by a boy of perhaps ten years. He was dressed in rags and bare-footed. "Buongiorno, Signore."

"And good morning to you, as well." Howell spoke to him in the Italian he had learnt working with sea captains in the counting house. He could not imagine that this ragmuffin knew any other language.

"You seem to be looking for something, Signore."

"Indeed, I am looking for an inn called 'Il Toro'. Do you know where that is?"

The boy's face brightened. "Oh, yes, Signore. I will happily show you the way for a decimo."

Howell extended his hand. "Content."

The boy tentatively shook Howell's hand as though no one had ever before made a bargain with him in this fashion. "Follow me."

They started up one of the streets leading off the jetty. It was not overly wide, but wide enought that there were vendors' carts at more or less regular intervals and shops with awnings scattered amongst them.

Suddenly Howell's guide stopped, grinning, and pointed to one side of the street. "It is a short cut to the inn, signore."

Howell took one more step so that he could get a good look down the alley before him. He knew what was coming. He had done it. The gentleman who had been lured into the trap would be set upon by ten or twelve boys. Some would wield makeshift clubs. Some would be armed with a bit of metal sharpened by grinding it against a stone. Many would only have their fists for the attack. They would beat the man senseless, and steal all that he possessed, perhaps even his clothing.

Howell smiled at his grinning guide, and slapped the boy across his face as hard as he could.

The boy went flying, and rattled to a stop on the cobbles of the street. He held his hand against the cheek that had been hit and stared, wide-eyed, at Howell.

Howell leaned over to menace the boy as he lectured him. "I know what you were about, and I won't have it." He drew his knife just enough that the urchin could see it. "Now, if you will really take me to 'il Toro', I will give you the decimo we have agreed upon."

The boy swallowed and, as he got to his feet, Howell seized his ear, digging into it with his nail. The urchin pointed back towards the jetty, and edged around Howell to lead them on their new journey. When they got back to where the ships were, Howell's guide took him up another of the streets. They walked around a very slight curve, and Howell immediately saw the sign for 'The Bull'. All he would have had to do was to start up the street from the ship, and he would have found it in an instant. Still and all, he had made an agreement with this boy and it was his duty to hold to it. He walked to the entrance, and he and the boy stopped.

The boy stood nervously, not sure what was going to happen next.

Howell let go of his ear.

The boy took off running without an instant's hesitation.

"Wait!" Howell cried. "Wait! I owe you a decimo."

The boy rolled to a stop, and looked back at Howell. He looked all around him, fearing that it was he who would be trapped. Then his eyes fixed upon Howell again.

Howell drew out his purse, and found a decimo. "This is the agreed price, I believe." He tossed it to his guide and potential betrayer. The boy snatched the coin out of the air, and ran off towards the jetty. Howell knew what it was to be that boy, and to be treated fairly. To be treated with respect. Howell had forced him to fulfill his part of the bargain, but, when he did so, Howell had fulfilled his.

Howell opened the door under a painted wooden sign of an angry bull, and walked into the public room of 'The Bull'. There were a few tables and chairs scattered about and some benches along the walls. The few patrons looked bored or drunk. The host got up listlessly from his place behind the bar.

"There is a man here who keeps an office for Riss and Grace." Howell stated it as a fact.

The host pointed at a door on the far side of the room, and sat back down.

Howell walked across the room, and removed his hat. He and rapped upon the door.

"Yes?"

"It is Howell Davidson, Sir. I arrived with the ship yesterday."

"Come in, Mr. Davidson. Come in." The man actually got the door open from the inside before Howell could open it. "Good to see you. Good to see you. Did you have any trouble finding the place?"

Howell considered discussing his little adventure, but he decided there was no point in upsetting a man he had only just met. "No, Sir. It was much as I'd expected."

This room had the usual plastered walls and white-washed ceiling, but the unusual thing was the large table set against one of the walls. Inns rarely had a side room furnished like this. Mr. Storey had arranged all his papers upon it in the rigidly organized fashion that Mr. Riss insisted that all his clerks maintain.

"A good place to do our business, eh?" Mr. Storey picked up a piece of paper, and started to read off a list he had prepared. "There is the matter of settling the fees we owe for the use of the harbor. I made the arrangements for the wagons and the teamsters we will need for our cargo before you arrived. Will you be certain that all the wagons have the necessary tools? Shovels, buckets, that sort of thing."

"Yes, Sir."

"We still need the means to transport all the supplies we will need, and this is all proving very difficult. People are reluctant to let their wagons and horses get out of their sight, what with all the trouble in Bohemia.

"Trouble, Sir?"

"Yes, of course. You wouldn't know. You were at sea when the news arrived. The Estates of Bohemia have declared their independence from the Emperor, and no one can tell where this will lead."

What did this news mean for him? Did it mean something to Sir Horace? Did this have something to do with the 'goal' that he suspected Sir Horace pursued?

There was no knock, but the door squeaked on its hinges just the same. It was held open by one of Sir Horace's men as the gentleman himself strode into the room. His blonde hair, twisted into curls by the damp air, emerged from under his hat. A hat that he pointedly did not remove in the presence of Storey and Howell.

"Had the most Devil of a time finding the place." He seemed to blame the agent for his difficulties. "Some street urchin offered to show me the way for a decimo. Can you imagine? A gutter rat like that? I agreed, and when we arrived, I told him to be on his way."

Howell's estimation of the man went down another notch.

Sir Horace gave the space a brief inspection. The man beside Howell was the agent of a respectable trading firm. A firm that he himself had engaged to transport his goods. Yet he looked at Storey and Howell with a frown upon his face. "These inns are barely above the status of a hovel."

Storey did not reply to Horace's insulting comment about the place of business he had found, but instead said, "How may I be of service, your Honor?"

"Yes. To business." Sir Horace pulled his portfolio from beneath his arm, and selected a sheaf of assorted papers. He swung them out towards Howell. He said to Storey, "See to it that these are properly dealt with."

"Certainly, Sir."

Howell took the papers.

"I trust that I shall find you competent in discharging your duties. Good day to you Storey, and..." *uhh* Sir Horace turned and left without finishing his sentence by saying Howell's name. Nor did he bother to close the door.

Howell watched Sir Horace and his man leave the inn, and then closed the door behind them.

Storey took the papers from Howell and put them on the table. "That seemed rather ill-mannered," he said to no one in particular.

"Fear not, Sir. You are not being singled out. Sir Horace is that rude to everyone."

"I see." He picked up Sir Horace's papers and began to read.

"With your permission Sir, I will go out and begin gathering the equipment that you have assigned to me."

"Certainly. Please do." He never turned to look at Howell.

Howell went out into the street, and turned back towards the docks. There he boarded the ship on which he had sailed from London, and spoke to the first sailor he encountered. "I need a crowbar."

The man looked at him as though Howell had asked him for the mainmast.

"Swiftly now. Fetch me a crowbar."

The sailor shook himself out of his stupefaction, and went off to find the required tool. He returned to Howell, and handed it to him.

"Thank you." Howell turned, and started off.

*Uhh* "You'll bring that back, won't you Sir?"

"What else would I do with it?" Howell replied, but he was already going down the gangplank. If he tried to use this crowbar to get one of the boxes open at night, he had have to carry a torch so that he could inspect the contents, and that would make him rather obvious. If he tried to do it in the daylight, he would certainly be seen by any guards that Sir Horace had posted. What he needed was a distraction, and he knew just where to find one.

He did not return up the street that lead to 'The Bull', but rather one further along the dock. He walked up this street until he found the alley that his 'guide' had tried to lure him into. He held the crowbar in one hand, and drew his knife with the other. He started in. He made no effort to conceal his entrance. On the contrary, he clanked the crowbar along the stone walls, and spoke loudly. "I know you're in there lads, and I want to hire you." He kept saying different things along the same line as he walked slowly forward.

Suddenly he was surrounded by ragged, dirty, armed urchins. Directly in front of him was a boy banishing a stick with a nail driven through the end to make a wicked club. This was the leader of the pack. Howell knew that this would be true from his days in

London's alleys. The boy with the most power would be the one to confront him. He had have to do it to keep the other boys' respect. He was about a head shorter than Howell. He was lean, bare-footed, his drawers in tatters, but he wore a nice doublet, doubtless recently stolen.

One of the boys to Howell's side raised his club to proceed as he had so many times before. Howell whipped around, and put his knife to the boy's throat. "Drop it."

The boy's club fell to the ground.

Howell displayed his knife to the multitude once more. "Who wants a bit of this?"

No one spoke, and no one moved. Howell shifted his attention to the leader. "Let us talk business." Howell kept his knife aimed directly at the leader's chest. He held the crowbar at a threatening angle.

The leader could not imagine what Howell was talking about. "Business? What business?"

"I need a group like yours. A group used to moving fast and confusing their enemies."

"And for what do you want my gruppo?"

"I want you to get some guards to look the wrong way."

The leader looked dubious. "And what's is in for me?"

"A quarter ducat." Howell was not a rich man, and this would be a considerable drain on his resources, but it was a fortune to boys who struggled to live on what they could find in other people's garbage, and what they could steal. The color drained from the faces around him. Boys looked at one another in astonishment. They would only get a fraction of what the leader kept for himself of course, but he could only keep his position if he dispersed some of what his gang took in.

Unlike the boys around him, however, the leader kept his composure, and he kept his eyes on Howell. His arm holding his club was cocked and ready. "So you will get us to do your work for you, and then you will disappear. No pay. No money. We're not fools."

The boys around Howell and the leader sniggered, and poked one another. They were starting to relax, though. Their clubs were not held at the ready.

Howell now relaxed as well, dropping his crowbar to a neutral position, and holding his knife across his body, pointing at nothing. He looked around at the dirty faces around him. "I was once like you. There was a time when I stole food from the markets to live. I worked as a cut purse, and lead strangers into ambush like this lad here tried to do." Howell looked at the boy with a big bruise on his face. "That is how I knew what you were about." He looked back to the leader. "So, here is the bargain I offer you. I will give you the quarter ducat when we reach the place that I will ask you to work. After that, you are free to honor our agreement, and distract the guards for me, or you may run off with my money and do nothing."

Howell and the leader both stood, and looked at one another in silence. One by one the other boys began to make little sounds urging the leader to accept. The leader relaxed the position of his club. Howell put his knife back into its scabbard on his belt. He dropped his crowbar until it was aimed at the ground.

"I can do what I want?"

"You'll have the money, and the choice of what to do after that."

The leader went back to thinking. Plainly, he had never been treated like this before. He had always been despised, kicked aside, known to be untrustworthy, the enemy of society. Now he was being offered the chance to act like an honorable man. It must have made his head spin. He hesitated.

"Silvio!" the boys around him clamored. "Take it!" "How can we go wrong?" "Per l'amor di dio, it is a quarter ducat!"

The leader shushed the boys around him. "What is this work? Where is it to be done?"

"There is a wagon park just outside the city." Howell pointed. "All you have to do is approach the wagons from the side I will show you, and start making a racket. Try to steal one of their purses. Throw stones. Yell curses at them. Anything to keep their attention until I have my work done. I will start as soon as I see you start."

"And how will we know when you're done?"

"I will walk a little way down the road towards the city. You'll see me toss this crowbar into the bushes." He held up the tool. Howell offered his hand. "Content?"

The leader hesitated.

"Silvio." It was the boy who had reluctantly guided Howell to the inn. "He paid me what he said he had pay me, just like he promised."

The leader smiled. He shook Howell's offered hand. "I will do it."

"Right. No mischief while we're on our way. We do not want anything to slow us down. Follow me lads."

It was dangerous for the whole pack to be seen together, so the boys darted here and there, always keeping either Howell or their leader in sight. Presently, they came to the edges of the city, and then Howell pointed out the wagon park. It was still quite some distance away, but Howell thought he could see two heads amongst the wagons.

Howell and his strange company stopped. He extracted his purse and pulled out several coins from it. The leader eagerly held out his hand. Howell pressed them into his palm, and the leader hefted them ever so slightly so the other boys could see the treasure he had found for them.

"Are you ready to do the work I asked you to do?"

"Si, Signore, si, si."

"I will circle around and come in on that side." Howell pointed to the end of the parked wagons farthest from the road that lead to 'The Bull' and the jetty where their ship was docked. "You come in on this side."

The leader surveyed the ground he was going to have to cross and the wagons before him. "Si." He turned to his crew, and began to give the other boys their instructions. Several of his boys looked scared or unsure. The leader looked at Howell out of the corner of his eye.

Howell started off to take his position. Was he a fool? Would the leader return to his instincts, sharpened over years of living on the streets, steal his money, and be off? Howell had given the boys far more than he could afford to have a peek into those boxes. Was it worth it? Would he get the chance to see what they contained?

Howell reached the best place for him to make his move. He waited. The guttersnipes had a shorter route, so they should have been able to start their attack by now. All Howell could do was hope.

"Give me that!" It was the voice of a man, not a boy. "Come back here, you little bastard!"

*Ha, ha, ha, ha.* That was definitely a boy's laughter.

Howell waited a moment until the commotion was well under way, and then he ventured to take a look. Two guards were dealing with the boys that Howell had paid, and Howell could see no other guards. The odd thing was that the leader of the street band had only two other boys with him. Maybe the courage of the rest of the pack had failed. Maybe the leader had told off the smaller boys to lay low. Whatever the explanation, the leader was living up to his part of the bargain. Howell was proud of the lad, as well as relieved that he could get on with his plan.

Howell darted towards the closest wagon. It was at an angle that would badly expose him if he got into it. He moved on to the next. It was tipped in such a way that it afforded him much better cover than the first. He hopped into the wagon bed. He jammed the crowbar under one of the boards across the top of a box, got down on his knees, and started levering off the board. It took a few tries, but then the nails came free and he could look inside. The box was filled with muskets and all the equipment needed to fire them. Then he looked closer and saw that the bottom of the box was covered in some dull gray bars. Lead! It was the lead needed to make musket balls. Thousands of musket balls. With all these wagons full of all these munitions, there was enough here to equip an army. There was one thing missing, however. There was no gunpowder. Howell had not seen any kegs of power unloaded from the ship, nor was there any in the boxes. What good were muskets without gun powder?

Howell beat the nails back into place as quietly and quickly as he could. He rolled out of the wagon, and hurried over to a place where he could be seen by the leader of the urchins.

"Hey! You! Stop!" yelled a man's voice.

Howell could not be sure if he had been spotted by one of the guards, or if this shout was just more of the fight with the boys.

He raised the crowbar high over his head, and gently threw it into the bushes.

# Chapter IV

## Destinations

Mr. Storey, the agent that Riss and Grace maintained in Genoa, looked over his table in the inn, and threw up his hands in despair. "What am I to do with all this?" The table was covered in neat piles of papers of every size from full-sized pages of fool's cap down to the tiniest scraps. Some had pins through the corners to hold them together.

"What can I do to help, Sir?"

"There is nothing that you can do, Mr. Davidson. That is the problem." He gave a big sigh. "The party that is going with Miss Kincade is bound over the Alps, that part is certain, but when? I have no idea where Sir Horace and his party intend to go or when. Do I prepare a place for him amongst those crossing the Alps, or does he want a separate journey elsewhere? Unless I can get some definite instructions as to where all of these people are bound, how can I prepare?"

Howell could see the problem. How many wagons? How many teamsters? When would they all be required?

"And as if this was not trouble enough, there's this Rebellion in Bohemia that's making it almost impossible to get horses and wagons."

A gloomy silence descended upon them. The bar room on the other side of the door was anything but silent. There was laughing and some gay shouting. At least someone had no doubt about what he should be doing.

The Mr. Storey muttered, "I cannot get a single word from Sir Horace."

The heat and stuffiness pushed themselves to the fore of Howell's mind now that there was nothing else to occupy it. What a ghastly climate. A stroll along the harbor jetty might be a trifle cooler. Suddenly a path forward struck Howell. "Mr. Storey, perhaps Miss Kincade could shed some light upon these questions."

"I already know everything that she can tell me that involves her travels."

"You misunderstand me, Sir." Howell was warming to his topic. "I am sure that you have gathered everything you need from her about her journey, but until she begins that journey, she is in contact with Sir Horace every day. They eat meals together. They use the same garden to relax. Who knows what she might have learnt about his plans?"

Storey nodded his head. "A capital thought, my dear Sir. Capital. Let me get my hat, and we shall be off."

"Mr. Storey, if I might intrude. There is no reason to suppose that she will be ready to talk the instant we appear. She may very likely prefer to talk where she cannot be heard by Sir Horace as well. Let me carry a note from you that asks for a meeting. If there is any standing about to be done, then I shall be the one who does it."

"I thank you, Mr. Davidson. That's very thoughtful of you." He turned to his carefully arranged desk, selected a bit of paper of a suitable size, and neatly wrote out a few lines. He signed it, and handed it to Howell. "I trust that this will get the desired response."

Howell put the note into his purse along with the few coins it still contained. He was not looking forward to a stroll through Genoa.

As disagreeable had been the air in the office, out on the street, in the direct Sun, it bordered on the unbearable. He turned towards the jetty, although it was not the most direct route to Rosy's lodgings. He found that it was a trifle cooler there.

Some of the ships stood idle, but these were a rarity. Most were being loaded or unloaded. Crates were moved in great nets made of rope the size of your wrist. Barrels were rolled along temporary ramps made of planks. Some ships were entering the harbor. Others were pushing back to put to sea once again.

He walked along the jetty as far as he could, and then he turned up the street. He arrived outside the impressive house in which Rosy was staying. He knocked on the front door.

The butler opened the door, and looked at him in a friendly way. He undoubtedly remembered seeing Howell escorting Rosy last night. "Yes, Mr. Davidson?"

Howell handed over the note. "Would you please give this to Miss Kincade?"

"Certainly, Sir. Will you be waiting for her reply?"

"Yes, I will."

"Well then, Sir, perhaps you would like to step inside and out of the Sun until she responds."

"Thank you." Howell stepped in, and removed his hat. It was stuffy in the antechamber of course, but it was miles better than standing in the street under the Sun. As his eyes adjusted, he looked about. Until this moment, the counting house of A. V. Riss and Grace had been the most luxurious place he had ever seen. To take in what was around him now was astonishing.

A door opened on the other side of the room. Sir Horace emerged. He was surprised at seeing Howell, but maintained a neutral expression. "Ah, Mr. Da… Dar…"

"Davidson, Sir."

"Yes, of course." Sir Horace considered Howell for a moment. "I do not recall summoning you."

"You did not, Sir. Mr. Riss's agent has sent me to collect Miss Kincade, so that they may discuss her coming journey." He knew he was being impertinent, but Howell could not resist continuing the conversation. "He has told me that he would like it if you would talk to him, as well Sir."

"Would he now? You may tell him that I am still making my preparations, and will speak with him at the proper time."

"Yes, Sir." Howell decided to press forward. He had already overstepped his bounds, so there could be little harm in going further. "Mr. Storey tells me that you may accompany Miss Kincade over the Alps."

"That is true."

"He also says that you are considering other places you may go."

"You are the curious one, aren't you?"

Sir Horace and Howell had fought the pirates together, and so they were brothers in arms, in a manner of speaking, yet Howell had seen him wound one of the ship's sailors in his zeal to kill one of the raiders. Howell would never have done that to someone who was fighting at his side. Rosy had said that Sir Horace kept to himself as long as she had know him, and Howell had seen that he could be quite rude. Yet here he was having a civil conversation with the man. Why? If Sir Horace was 'driven' by something, Howell saw no evidence of what that might be. Howell wanted to understand this fellow.

"Good day to you, young man," Sir Horace said, and took his hat. The butler opened the door for him, and Sir Horace went out into the street.

Howell was left in the vestibule, thinking about what he had learnt and not learnt from this latest encounter with Sir Horace. All the mystery that surrounded this man was almost certainly connected to the cache of muskets and lead that Howell had seen. It was all very thought provoking.

"Would you care to sit, Sir?" It was the butler speaking to him. His tone was not groveling, but respectful. He relied upon his wages, as did Howell, so there was a certain common ground between them.

"Yes, thank you." Howell sat, and waited for Rosy, and sweated. This was Summer, however, so perhaps it was more bearable when the season changed. He looked up at the butler. The man was looking blankly into the air. He did not want to intrude into his guest's affairs by looking at him, so he was leaving Howell in peace.

There was a knock at the front door. The butler turned, opened it, and looked out. Howell caught just the briefest glance of a face beyond the butler's knee. It was a terribly disfigured face. The butler stepped out, and closed the door behind him. Almost immediately, Howell could hear the *clank* of a coin being dropped into a tin cup. A voice said something, and the butler replied. Then the butler came back inside the house. Howell realized that the butler had drawn upon his own meager collection of pence to help a beggar in terrible need.

Howell got up from his seat and stretched.

"Can I do something for you, Sir?" said the butler.

"You have already done something for me."

"I beg your pardon, Sir?"

"You have taught me a lesson that everyone should learn."

The butler just stared blankly.

Howell took up his purse, fetched out a coin, and handed it to the butler.

"What is this?" The butler realized that he was speaking with far more familiarity than was appropriate. He immediately corrected himself. "I beg your pardon, Sir. What is it that you want me to do with this coin?" Doubtless he expected that Howell would tell him to go into the street and buy something to eat, or something along that line.

"I want you to keep it."

The butler looked at Howell with doubt in his eyes. "Keep it, Sir?"

"I saw what you did for that unfortunate out in the street just now. I am sure that you have little enough to spare in that purse of yours, yet you shared what you had. That should be recognized. That should be rewarded."

"Oh, Sir. No, no. What I did was just something I wanted to do for that poor man." He held out the coin to Howell.

"And I am just doing something I want to do for an honorable man."

Solemnly, they stood looking at one another. Then Howell suddenly reached out, and gave the man a friendly shove on his shoulder. They both broke into great smiles, and looked at one another again. This time their gaze was in joy.

Howell sat to continue his wait. Now he and the butler looked at one another occasionally, and smiled in satisfaction. It almost made Howell forget the heat for instants at a time. At last, the door swung open, and Rosy stepped into the room with an attendant by her side. Rosy must have been feeling the same heat that Howell did, but it did not seem to affect her at all. She looked as beautiful and confident as when Howell had first seen her in London. She had not made the slightest effort to apply some sort of face paint to hide her freckles.

"Howell," said enthusiastically. Then she glanced at the butler, and hastened to correct herself. "Good day, Mr. Davidson. You are looking well." The butler was smiling. Rosy had no complaint about this, she was perfectly content that the man should be happy, but it was just so inexplicable. She had never before seen him except with a neutral, dutiful expression.

Howell put her hand into his arm. "He and I have become quite friendly while I awaited you."

"Oh, goodness. Was I really that long?"

"It was actually a very great pleasure, Miss Kincade." Howell glanced at the butler, and then back at Rosy. "Shall we begin?"

"Certainly." She nodded that her attendant should stay behind.

The butler opened the door for them, and they stepped out into the street. The heat was every bit as disagreeable as it had been when Howell had left the office. Now, however, he had a beautiful woman accompanying him. Last night he had taken her on an adventure the likes of which she had never imagined before. He had shown her a world without gossip, without rules, and without restraint. They had shared out the money that they had made by taking advantage of the drunks that Rosy's beauty lured, and they had done it using stolen goods. They had kissed, and Rosy showed every sign that she was not done with adventuring yet. Where would all this lead?

Besides these recollections, their time together gave him the opportunity to discuss this Sir Horace who nagged him so. "I have been looking into the affairs of our dear friend, Sir Horace."

Rosy gave a chuckle. "Have you now? And what have you found, pray tell?"

"Well, Sir Horace has a few secrets I am sure he had like to hide."

Rosy became solemn. "I too have found that he is very fond of secrets, but I feel that you are talking about something more serious than hiding tobacco."

"Indeed." Howell steered them around a wagon loaded with hay. "Do you recall that we began to unload directly we pulled up to the pier?"

"Yes. It was all very prompt," she said.

"The first of the cargo that was pulled from our hold belonged to him, and it was placed into his wagons as quickly as it came out."

"Yes." Rosy reflected upon this. "When we were on our voyage, he insisted that his cargo was 'nothing special' and merely 'the sort of things that ships carry.' Yet when we arrived here, he seemed to take a particular interest in seeing that his boxes come straight off."

"Yes he did. He watched as every one of those enormously heavy crates came out, even as I was doing the count. I have found that there is an interesting reason for that."

"Yes?"

"I have managed to break open one of his crates, and see inside. The top layer in the crate was made up of muskets, bandoliers, shot bags, bullet molds, and everything else a musketeer might require. The bottom of the crate was covered by slab after slab of lead, enough to make thousands of musket balls."

"Whatever does he want with all that?"

Howell nodded, thinking. "I wish I knew."

"And why do you care?"

Howell thought about that, as well. Why did he care? What about this situation drove him to search for Sir Horace's story? "I saw him drive his rapier through one of his shipmates in order that he might stab a pirate. He showed no reaction at all to what he had done. Even when the fighting was ended, he made no effort to find that man he had wounded, and help him. He had done what he wanted for himself alone. I want to know who this man is. I want to know what devilment he may be about."

Rosy absorbed this. "I have seen the unfeeling way in which he treats the staff. I have seen him totally focused on his own business, but I have never suspected he could be ruthless on such a scale. What might he want with all that military equipment?"

Howell spoke with the memory of what he had seen in Sir Horace's crate. "There are at least two things that I can think of. The Emperor is making loud noises about converting everyone in the Empire into good, obedient Catholics, but he does not have anything that you could really call an army. Sir Horace might be thinking of using these arms to help the Emperor to raise and arm troops to enforce his religious wishes. In the process, of course, Sir Horace would make the Emperor indebted to him."

"Yes. And the other?"

"The other possibility is the exact opposite. Resisting the Emperor's pressure are quite a few Protestants in the Empire, the Rebels in Bohemia being foremost amongst them. Horace may be thinking of supplying these, and gaining influence in the process." Howell looked at Rosy significantly. "There is one more factor we must balance on our scale. Amongst all the musketeer kit, there was no gunpowder. Muskets without powder are no more useful than just so many sticks."

The couple found a shady spot, and stepped in. There was no breeze, so it continued stifling hot and smelling of the corruption in the streets, but at least the Sun did not beat down upon them. Rosy seemed indifferent to what they were enduring, and she continued as though nothing was amiss. "If Sir Horace intends to arm the Protestants, I know exactly where he is going."

"Exactly?" How could anyone be so certain?

"The Protestants on the other side of the Alps cannot agree on what they should do, but they all agree that they do not want the Emperor meddling in their affairs. If Sir Horace's intention is to arm them, then the man he will approach is the husband of my good friend, Elizabeth. He is Fredrick, the Count of the Palatinate. He also is the nearest thing that the Protestants have to a leader."

Howell mulled this for a moment. "Horace is from England, so he is a Protestant, is he not?"

"I doubt that anything as trivial as his immortal soul will guide Horace's choice. He will sell just as readily to His Most Catholic Majesty as to the Protestants, considering only his own advantage."

Howell had no trouble believing this. Howell tried to put all these pieces together. Sir Horace could use his armory to support the Protestants, but he would gain influence in over a loose collection of little dukedoms and principalities whose members were often pulling in different directions. He could gain influence with the Emperor, but once he had handed over his arms, the Emperor might consider Sir Horace's work was done, and dismiss him. "Rosy, do not put yourself in a position where Horace might suspect that you know anything about his plans."

"How can I promise that? We are lodged under the same roof until we can begin our journey over the mountains. I might hear anything."

"You will not hear anything about this. I guarantee that these weapons are the most valuable bargaining chip he's ever had, and that he will carefully guard his intentions."

She looked at him with determination, a prospect of adventure in her eyes. "Perhaps I should discover these intentions."

"This is a dangerous piece of business," he said. "Doubtless Sir Horace sees this as his chance to win great power, one way or another, and he will let no one survive that threatens to interfere with that objective."

"We shall know more presently." Rosy spoke earnestly. "You are taking me to visit with Mr. Storey, are you not? Our company is bound for Fredrick's court beyond the Alps. He has the preparations in progress now. Will Sir Horace accompany us, or will he stay in Italy, do you think?"

"That is Mr. Storey's problem as well. He is hoping that you will be able to shed some light on all this confusion."

Rosy smiled. "I am afraid that Mr. Storey will be disappointed. I have not heard anything to discover Sir Horace's intentions."

When Horace made his choice though, it would certainly give Howell another clue to fit into the riddle he struggled to unravel.

They found their way through the busy and steaming streets of Genoa. Under their awnings, shopkeepers sold freshly baked buns with their aroma buried in the stink everywhere around them. Howell and Rosy had to stand aside as carts, and once even a flock of sheep, pushed by them. They arrived at the door of 'The Bull' and stepped in. There was the same listless crowd in the main room that Howell had seen before. He knocked on Mr. Storey's door.

"Come."

Howell ushered Rosy into the presence of Mr. Storey and his perfectly organized desk.

Although Storey had issued the invitation, he seemed to be taken off guard by Rosy's sudden arrival. "Oh. Yes. Good day. Thank you for coming, Miss Kincade. Yes." *mmm* He looked around for a place where she could sit. He had nothing that he could offer.

"Shall I find a chair for Miss Kincade, Sir?"

"Yes, Mr. Davidson, that would serve wonderfully." Storey was himself again.

Howell went into the public room. There was the usual crew of idlers sitting about. Some sat in morose silence. Some chucked their mates in the arms, and laughed at jokes only they could understand. There was plenty of room on the benches, but all of the chairs were occupied. Howell walked up to the one nearest to Storey's door. "I'd be obliged if you would give me this chair. I need it for a lady."

The man turned around to look at Howell over his shoulder. "You do not say?" He turned away.

"Oh, but I do." Howell walked around the man. He leaned over to put his nose just inches away from the nose of the man holding his mug. "Perhaps I did not make myself clear. I need this for a lady. Here is your chance to act the part of a gentleman." Howell continued to loom over the man.

The room fell silent.

"The benches seem inviting."

"Yes." The man got up. "Yes they do." He wasted no time in going across the room.

Howell picked up the chair, turned his back on everyone, and went into Mr. Storey's office.

"Ah, Mr. Davidson." Storey pointed to the place he had selected. "Miss Kincade, would you care to sit?"

"Yes, thank you." Rosy glanced at Howell, not sure of what all the commotion in the bar room had to do with the chair. She never paused, though. She stepped to where Howell had placed the chair, and took her seat.

"Miss Kincade has been telling me that Sir Horace has not revealed to her anything of his plans. This is very unfortunate, do not you think?" It did not matter to Storey whether or not Horace decided to go with Rosy and her company over the Alps. Storey was only concerned about getting the necessary wagon train assembled at the time it was needed.

This was not true for Howell. He wanted to watch this business with Horace play out. More importantly though, this man might present a threat to Rosy, and Howell wanted to be certain that she was safe. Howell might be able to know everything that was going on, and that required that Mr. Storey keep him informed. There was no guaranty that he would do that, however. Storey wouldn't deliberately keep him in the dark, but he just might take some action that was part of his job, and Howell not know about it. However, there was another way that Howell could learn about Horace's plans. He could reveal what he knew about Horace's cargo, and hope that Storey shared his concern about what those arms represented. Howell hoped that this risky move would give him what he needed. "I know when Sir Horace will tell you what you need to know to make your preparations."

Rosy looked at him in confusion.

"Ah, what a relief." Storey was all smiles. He went back to his desk, took a piece of paper, and held his quill poised above it. "Yes, Mr. Davidson?'

"He has four options, each of them with unpleasant results. He will move when he has decided which plan he has decided to execute."

Mr. Storey straightened up, his quill still in his hand. "Unpleasant plans?"

"Sir Horace is not just a self-centered ass. He has the means to make himself one of the most powerful men in Germany. He has the means to change the course of all of the affairs unfolding about us even as we speak. His wagons? They are filled with the goods necessary to equip all the musketeers of an army."

"Army?"

"Muskets. Bandoliers. Lead for musket balls. He may carry all this to the Emperor. He may take it over yonder mountains to the Protestants. Once he's made his delivery, he will have gained the unquestioning trust of those he delivered to. This gives him two more choices open to him. He can remain loyal, or he can use his new position to spy for the other side."

Mr. Storey was speechless.

"That is why we…" and Howell indicated Rosy "…need your help."

Storey gave a great sigh. He looked carefully at Howell's face. "Help." He did not say it as a question, but rather as though he was considering the word.

"It may be a much as Miss Kincade's life is worth."

Storey glanced at Rosy in alarm, and then back at Howell. "What would you have me do?"

# Chapter V

## Sir Horace's Plans

Sir Horace trudged through the foul streets of this ghastly port. The Sun beat down upon him. How could anyone survive in such a climate? It was like a foretaste of the Hell that threatened them all. He saw people in the distance who shimmered and writhed about. They might be damned souls doomed to be tortured for all eternity. Thank God his lodgings were just ahead. At least there would shade there, and he would be able to cool sufficiently that he could think clearly. He mopped his brow.

When he reached the door of the building that he shared with that harlot, Miss Kincade, and all her their entourage, a butler was there to open the door. Sir Horace pushed the man aside, and shoved his hat and portfolio into the servant's hands. Finally. Finally out of that blazing Sun. He pointed at his folder. "Bring that and some water to me in the garden." He started off for where there was shade and trellises wound with vines and calming shrubs surrounding it all. He needed a place that was cool and soothing where he could ponder some recent developments.

That reckless fool who had fought the pirates, what was his name? Davis? It did not matter what his name was. This madman had been seen by one of his guards leaving his wagon park. Horace could not be certain what, if anything, this meddling boy had seen, but he had to assume that knowledge of his weapons was now known to far more people than he would have wanted.

There was other news as well, and this was known to everyone in the city. A German Duke was in Genoa on an official visit, and was almost certainly making some sort of arrangement with his host, the Doge. This would doubtless serve to strengthen the Protestants. This gave Horace a possible course of action once he had crossed the Alps.

The butler arrived with a glass of water with a wedge of some sort of fruit on the rim. Horace shooed him off. He swallowed a deep draft of the slightly cool, slightly fruity water. He had to come to a conclusion and act. He was haunted by the knowledge that every moment he tarried narrowed the options available to him.

His most obvious course of action would be to accompany that red-headed bitch on her trip over the Alps to visit her long-time friend, Elizabeth. This woman was the wife of that dunce Count Fredrick, who thought he spoke for all the Protestants. Horace's weapons would be an effective attention-grabbing gift to him. Horace was confident that Fredrick would be easy to bend to his will once the muskets and whatnot had been handed over. Horace could almost certainly become the power behind the throne of this weak-willed Count.

Another choice was to hand over his weapons to the Emperor. His Majesty was trying to gather troops to put down the Rebellion, but was making only slow progress. Once on the far side of the Alps, it would be a simple matter to leave the others to finish their journey, and take his wagons in a different direction. He could then rendezvous with what little there was of an Imperial Army. In exchange for his arms, Horace could ask for any number of rewards. He might ask for a considerable tract of land somewhere in the Empire. He might ask for a prestigious command in the Emperor's army with plenty of opportunities to demand contributions from the various lords he encountered.

His glass was empty and he felt like a kettle boiled dry. "Boy!" Where was that witless servitor? He was never about when one required him. "Boy!" Sir Horace now lost what little patience was left to him by this climate and the mysteries with which he wrestled. He hurled his glass into the house, where it shattered.

The butler stepped into the doorway.

"Clear up that mess and bring me more water. No. Bring me more water and *then* clear up."

The butler disappeared.

He felt that he was in the hands of the Fates of ancient Greece. Was he locked into a pre-determined destiny, or could he act on his own?

"Here you are, Sir."

He took a sip of the water.

Now he was beginning to see possibilities beyond the obvious ones. There was a sinister twist he could put upon any approach he made to the Emperor. Once he had become a member of the Emperor's inner circle, he would be in a perfect place to learn everything that the Rebels would like to know. He could demand a high price for information like this.

There was a good reason to reject that course of action, though. Fredrick was a dunderpate, and all the Protestant Lords that he thought he was leading were pulling in different directions. Relying on such a shaky alliance was just too dangerous.

It was the perfect situation for his best option of all, though. He would present his arms to Fredrick and become the hero of the hour. Horace would be able to stay at Fredrick's side indefinitely. He would make himself the Count's most trusted advisor, and learn every decision and move that the Protestants made. He would then transmit everything he learned to the Emperor. He could ask the Emperor for the same rewards that he had imagined before. He would make sure that Fredrick never realized that the absence of gunpowder made him effectively defenseless. That would please the Emperor still more.

There was yet another step he could take to enhance his chances of success. He would make the Protestants feel even more threatened than they did now. He would attempt to assassinate the visiting German Duke who would be in the parade tomorrow.

Whether the attack succeeded or failed was unimportant. It would panic the Protestants whichever way is ended. Then Horace could plant any seeds he wished in any of the minds that he wanted to entangle in his vines. Ideally, the 'assassin' would escape, only to be killed soon after by one of Horace's men. This would ensure that no one could connect him to the assassination attempt. The news of Horace's quick action would improve his standing with the Protestants even before he began his trip over the Alps. All the Protestant Lords would fear that they might be the next victim without the protection that Horace could provide. Horace knew exactly who would play the role of the 'assassin' in the little drama he planned.

With that fool Fredrick in his grasp, Horace could offer to kidnap Count Fredrick and Countess Elizabeth, and then deliver them into the hands of the Emperor. These were some of the leaders of the party that sought to destroy 'the public peace'. How could the Emperor pass on the opportunity to execute such criminals in public?

"Boy!"

The butler stepped into the garden. "Yes Sir?"

"Bring me paper and ink."

"Yes Sir." The butler turned, and went back into the villa.

Four plans, eh? Four chances to gain undreamt of power. Soon he would be the power behind the throne of one party or another, and the trusted agent of another.

The butler returned and laid out the writing implements.

Sir Horace began to scratch out the first message he was going to send. "To His most Catholic Majesty…"

# Chapter VI

## Friend or Foe?

Howell and the butler exchanged greetings as Howell entered Rosy's lodgings. The butler cleared his throat in the approved fashion for him to get the attention of a superior. "Sir Horace has asked that you await him in the garden when you arrived, Sir." The butler seemed as unsure about this as was Howell. "I have been asked to bring Miss Kincade there as well."

Howell found a shady spot, and sat. He made sure that he left enough room that Rosy would have a place to sit beside him. The butler brought him some water. Howell wanted to just relax and enjoy the garden, but he could not stop thinking about why Sir Horace should take such trouble to make the contents of his wagons a secret? Well, this is Italy. Everyone he saw around him was a Catholic, and might carry the news to the Emperor. If Horace intended to take his arms across the mountains to Count Fredrick and the Protestants, then he wouldn't want it to be common knowledge. Was that the reason, or did Horace have some other scheme?

Rosy stepped into the garden. The butler pointed out Howell and then withdrew.

Howell stood. "Rosy. Good day." He held out his hand.

"And to you." Rosy looked over her shoulder as she approached Howell to be certain that Sir Horace had not appeared

Howell pulled her close. They embraced. They kissed.

"Howell. I wish we never have to be parted again."

Howell did not want to be parted from her either, but his place in the world was far too unpredictable for him to hold on to much hope that he would be able to manage it.

Sir Horace came through the door into the garden with an uncharacteristic smile on his face. "Ah, here we are."

Rosy disengaged, a trifle embarrassed of being caught in this position.

Howell gave a little bow. "Good morrow to you, Sir Horace."

"And to you, Miss Kincade." He turned his attention to Howell. "I do not know what manner of paper shuffling that agent of yours has in mind for you today young man, but I know of a better entertainment than making entries in a ledger book."

"Yes Sir?"

"Then let us be off."

Rosy was a trifle angry at being left out of this 'entertainment'. "I will have my lady bring my bonnet, and then we can be away."

Sir Horace was surprised by this, and was none too happy about it. He stood, and tapped his foot as they waited. Rosy stood serene at her triumph in making Sir Horace wait.

Howell thought this over. Sir Horace certainly enjoyed his little secrets, did not he? Where was he taking them?

At last Rosy had her bonnet, Howell and Horace collected their hats, and they stepped out into the roasting streets. Horace took Howell and Rosy in tow, and started for the center of town. "The Doge is welcoming the German Duke who is currently visiting. I expect there will be quite the cavalcade to show off his honored guest."

Howell looked at Rosy. "I'd like to see that."

"I would, too." She settled happily into walking along with Sir Horace, and her arm within Howell's. "This Duke is a rather prominent figure amongst the Protestants who oppose the Emperor, I believe."

Howell was constantly amazed by this woman. She spent her time in the court with the Countess, so it made sense that she would know something these things. It was very impressive all the same that she should know so much about what was ordinarily considered the business of men.

"Yes, of course." Sir Horace seemed unpleasantly surprised rather than impressed by Rosy's knowledge.

Rosy continued her analysis of what they were about to see. "The Doge has already promised to dispatch troops, Mansfeld's Regiment I believe, over the Alps, while loudly proclaiming that these soldiers demonstrate his support of the Bohemian revolt."

"Indeed," said Sir Horace.

"Now we find a German Duke who's conspicuous amongst the Protestants who are helping the Bohemians. We find him here hobnobbing with another potential supporter." She looked at Howell and Sir Horace. "Quite interesting, is it not?"

Sir Horace continued to lead them, and gave Rosy a little smile. He ignored Rosy's observations of the political scene. "It should be quite a show. The Doge likes to display his wealth, and what better way than by upstaging a man of equal rank?"

Howell thought that Sir Horace could fit that mold as well.

In a moment they came to a row of people along the margins of a major street lining the parade route. These people were every sort of citizen who lived in Genoa. The elite sat in their carriages so as to remain out of the Sun. Some spectators were respectable burghers and their wives. They stood beside their well-mannered children, who were supervised by a maid. Cobblers and butchers and other such tradesmen chatted amongst themselves with no concern for their unruly brood. Harlots sought customers. Beggars crept amongst them all, seeking alms from anyone who would take pity upon them. There were, inevitably, boys with theft on their minds. Defining the front margin of the parade route was a long line of civil guards. For this gala, they wore yellow and black checkered ceremonial clothing. They were armed with spears with both a point and a small axe blade at the tip.

Howell and his small party found a place for themselves amongst the onlookers. The Sun rose and Howell sweated in his heavy clothing. He had not bought new clothes because the intention had always been that they should go over the Alps, and Howell knew enough to expect that a mountain pass was likely to be a rather cold place.

"This is rather nice, do not you think?" Rosy patted his arm. "When Elizabeth goes out for something like this, I have always been back at the palace seeing to the preparations for the reception." She smiled up at Howell. "I like being able to see it from this side."

Howell tried to think of it in those terms. He had never worked on a crew that was putting together a banquet, never mind overseen one. The biggest meal he had ever had was a leg of mutton and a bottle of brandy that Mr. Grace had brought back to the counting house. It was a feast, and to be sure, but everyone that Mr. Grace knew was there, and there was nothing grand about it. Everyone had just tucked into whatever they could reach.

The sweat ran out from under his hat and down his face. He felt very out of place here. It was not just the blistering heat, although that was certainly a part of it. The buildings looked different. The smells were different. The smells of the food, certainly, but even the sea smelt different from what he was used to. Sometimes he felt like the people around him were staring at this odd stranger amongst them.

At last cheering could be heard coming from the lines of people down the street from where they were waiting. Group by group, the cheering was picked up by those who had come to see the show. As it came closer, one could hear iron horseshoes clattering upon the cobbles. Four horses appeared, riding abreast. These stern-looking fellows were guards, protecting the quality who followed them. Each wore a breastplate. Each carried his sword already drawn, vertical, as though it was growing out of his thigh. They did not smile, but rather kept scanning the crowd, looking for any potential trouble.

Behind them, two horses appeared. These men were dressed for an entirely different sort of business. The lavishness of their clothes was beyond measure. Each wore a sash of a brilliant color that clashed with that of his doublet. The lower margins of these were decorated with a fringe of gold, and even the button holes of their doublets were embroidered with the gold thread. Short, striped sleeves covered the tops of their bright white shirts, blinding in the full Sun. The color of their breeches contrasted with that of the rest of their costume. They wore spurs that looked like jewelry, and the tops of their boots were turned down to reveal yet another color. Their hats were decorated with enormous plumes. Even their horses were decked out in a fashion richer than most men would enjoy in their lives.

The Doge was riding slightly behind, letting his guest take the place of honor. The German Duke waved graciously as he rode towards Howell. The cheering was all around them. One of the titans of business waved a kerchief from the window of his carriage as the Duke passed right in front of Howell.

Suddenly an arm shot out next to Howell's ear. He had a brief glimpse of a hand holding a pistol, and then sparks flew from the ignition. *BANG!* The pistol discharged in the direction of the Doge and the Duke. Howell saw the man drop his pistol, and then Howell turned to look through the gun powder smoke to see the gunman force his way out through the crowd.

No one knew what had happened, least of all Howell. Everyone instinctively pulled back, leaving a void in the crowd just were Howell stood. Some were shocked to silence. Most yelled, and struggled as they tried to push back the people who were behind them.

Howell's first reaction was to chase the shooter. This criminal needed to be captured! Howell followed the man, forcing his way through the tangle of onlookers. The gunman was doing the same thing right in front of him. Then another thought came to him. Hardly anyone had seen the real assassin flee, but everyone was now looking at him.

"He shot!" someone yelled pointing at Howell. "He's the one who shot at the Doge!" Howell was seized from every direction. He twisted away from the first man to lay hands upon him, and immediately drove his fist into the jaw of another man who was blocking his escape. This fellow staggered to the side, and Howell shoved him mightily to ensure that he fell, and blocked the access of several more of the men attacking him.

"Howell!" It was Rosy, almost drowning in the mass of people around him. "Howell!"

The riot around him grew in intensity. Howell punched, and thrust his way through the mass of men and women in the ranks closest to the parade route. Then he was past them. The people he passed now were the furthest from the road, and had no idea what was going on. He was in an open space. He sprinted away from the scene. The citizens he passed simply watched him fly by.

Now came the guards who had lined the route in their bright party clothes. "Clear the path!" Genoa's civil guards forced their way after him. "In the name of the Doge, make way!" One of them swept his arm to signal his soldiers. "Comrades, make haste!"

Howell looked about, and chose the route that seemed the best available in his quick inspection. It was a passably large street, but he felt confident that it would lead to an alley where he could use his old skills to evade capture.

"Stop! Stop in the name of the Doge!"

Howell did not find the authority of the Doge to be all that compelling just now. He ran across the sunlit square and into the shady street. Vendors looked at him in confusion as he dashed by. There! He ducked into an alley, and never broke stride. He dodged around barrels, shovels, stacks of fire wood, and heaps of uncollected trash and animal droppings. He could hear his pursuers starting into the passage behind him.

He emerged in another street, turned, and aimed at Rosy's residence without hesitation. He dropped back to a brisk walk, hoping that he wouldn't attract attention. Everywhere he looked, people were buzzing about what had just happened. Howell caught snatches of this talk as he went by, and plainly there were more rumors than facts. "…the Doge's horsemen…" "…they were all killed…" "…blood…"

He came to another street that he knew led to where Rosy and her household, including Sir Horace, were lodged. He and Sir Horace had fought the pirates together. They had stood as one in the defense of the ship. Despite their differences, he hoped that Sir Horace would feel honor-bound to stand by a comrade. The one place in Genoa that he could be certain was a safe haven was in Rosy's lodgings. Howell moved quickly, but as soon as he turned the last corner, he was confronted by a group of the Doge's guards. He rolled to an easy pace.

The soldiers paid him no mind, and just continued to chat amongst themselves. He passed by them casually, and walked up the stairs to where Rosy had her rooms. He knocked.

The butler opened the door. "Oh my goodness. Thank God you are safe, Mr. Davidson. You know..."

"Yes, I do. Where is Rosy, that is to say, Miss Kincade?" Then Howell noticed that Sir Horace was already in the entrance hall.

"Ah, the inquisitive young clerk." Sir Horace appeared to have just arrived. "You have heard, of course, that there has been an attempt to assassinate the Duke?"

"Yes. They think I did it."

"You?" Sir Horace was stunned. "We must deal with this immediately." Sir Horace looked at Howell intensely. "You have stood by my side when all around me was danger and death. Can I do less for you?" He thought for another moment. "This would seem to be a place of safety, but the Doge's men will soon think along those same lines. It is the logical place for you to seek refuge. We must assume that they will soon come to the same conclusion, and come here to look for you."

Howell thought about this. Would the guards really force their way into Rosy's residence? Might they camp out, waiting for him to emerge in another attempt to escape?

"We must foil them." Horace smiled at how his plan would serve the needs of the moment. "We shall leave immediately, before the guards have worked out that this is the place upon which to focus their attentions. Come." He led Howell out the front door.

Contrary to what Howell had hoped, the guards had not left, but continued to stand across the street from the entrance to Rosy's lodgings. If anything, there were now more of them than had been there before. It appeared that someone might be thinking along the lines that Sir Horace had feared. The soldiers looked very closely at Howell and Horace, but made no move. Howell merely glanced at the guards, showing no interest in them. He simply ignored them for the most part as he walked by.

Horace laughed loudly as though he and Howell were sharing a joke. "We will be past them in a moment." *Ha ha ha!* His words and his actions clashed outrageously, but this was exactly what the moment called for.

Howell grinned and sniggered, but under his breath he said, "Some of them are looking at us."

Sir Horace wiped his eyes as though trying to recover from Howell's hilarious joke. "Pay them no mind."

They walked around a corner, and started down another street. Howell tried to press their pace.

Sir Horace hung back. "Calm yourself, Sir. Calm. We must appear to be ordinary visitors to this great port. No one worth giving a second glance."

Howell accepted the advice. He stopped trying to push their speed and slowed down enough to walk quietly beside Sir Horace. He could not stop thinking about his situation, though. A considerable street crossed their path just ahead. Which way would they go when they got there? Hell, even if he escaped the city, what would he do then?

They approached an alley, and one of Horace's guards leapt out. He wrapped his arms around Howell, pinning his arms to his sides.

"Take him!" Sir Horace yelled up the street towards the civil guards. He pulled his rapier. "Here is the man you seek!"

God's blood! Betrayed by this treacherous snake.

Howell twisted in the stifling grasp and tried to keep his attacker directly behind him, but it was all for naught. Horace plunged his needle-pointed blade into Howell's side, where it scraped along a rib. He pulled it back for another go.

He lost his chance to stab Howell a second time as the civil guards who had been standing in front of Rosy's door arrived. The guards in their yellow and black checkered clothes swarmed around him.

Howell got the briefest glimpse of the man that Horace had set upon him. Howell had a dark spot upon his doublet, and blood upon his hand. Horace had shown no hesitation in stabbing his own man just as he had done to the poor bastard on the ship.

He should have seen that his 'chance' meeting with Horace in the lobby of Rosy's residence was just too pat. Howell had thought he had worked out what Sir Horace was about. He had thought that Horace had sufficient honor to help a man who had stood by him in battle. Instead, Howell had fallen directly into the trap that Sir Whore's Ass had set for him. Howell had been dim enough to trust this villain, and now he had been served up on a platter. This man would pay. Howell would gut this hog. Howell would overcome whatever obstacles might be put in his way, but he would kill Sir Whore's Ass.

The officer strode up behind his guards, and fixed Howell with an evil grin. "He must answer for his crime." He looked around at his men. "Take him to the cells." This capture would be quite the feather in his cap. He almost laughed. The guards tied Howell's hands behind him, and did not scruple to pull the knots as tightly as they could. His wound was ignored. A man on each side of him grabbed his arms and he was led, or rather hauled, away.

Howell glanced over his shoulder. Sir Horace, that treacherous turd! Sir Whore's Ass, who would sell him, nay kill him, to get whatever it was that he wanted. That rat was going to pay for this. Howell would not rest until he had had his revenge. For an instant Howell was blinded by his rage, but then he returned to the moment.

He had a far more immediate problem. He was in the hands of men who thought he had taken a shot at either their Doge or his honored guest. Perhaps he would get a chance to deny that he had attempted to assassinate one or the other of these exalted persons. More likely they would not bother with any sort of legal proceedings, and just hang him out of hand. He was surrounded by a dozen or more guards. His hands were tied. Despite his most earnest thinking, he could not imagine a way to get himself free.

The guards encouraged him to keep up the pace with the points of their spears. As they marched along the streets, the officer called out, "This is the man who would kill our Doge!" People along the route spat upon Howell, and covered him with the muck of the streets.

Presently, Howell and his escorts came to the gates of an impressive fortress. Unlike the rest of the whitewashed city, this castle presented walls of raw, grey stone. He was taken over the drawbridge and into the courtyard of the fortress.

The officer of Howell's captors held his nose against the stink coming off Howell. "This is the one who tried to kill the Doge," he said to the first guard they encountered. The man spat upon Howell. The fellows with their hands around Howell's arms threw him to the ground. Everyone who could get a clear shot kicked Howell in his legs, his ribs, and even his head.

Howell was taken into the custody of the prison guards who dragged him to his feet. The bright courtyard gave way to the gloomy guards' room and then to an equally depressing passageway. At the back of this was a stairway, and Howell was taken down into the clinging darkness.

# Chapter VII

## Giovanni del Merico

Welcome! Welcome!" Partway down the row of cells was a ragged peasant smiling at Howell.

The place stunk of shitte, of course, just like Howell smelled from his encounter with the citizens of the city. There was the musty smell of mildew. There was something else here, as well, something that Howell could not place. Perhaps it was decay. Perhaps it was death. He looked down the aisle between the cells. There were no walls between the cells, just more bars. It was like a row of cages for animals.

"Put him here," said this cheerful prisoner to the guard, pointing to the cell beside him. "I could use the company."

The only company that Howell wanted was Sir Whore's Ass. He wanted Sir Whore's Ass standing beside him, and Howell with a knife in his hand. He with Sir Whore's Ass lying dead at his feet.

The guard unlocked the door of the cell opposite the happy inmate, and shoved in Howell. "How 'bout I put him in here?" He used his most sarcastic tone. "Den you can see him whilst you chat."

The smile on Howell's companion never wavered. "Fine. Fine." He looked again at Howell. "Welcome."

"Yah, I am sure as shitte dat he's over-fucking-joyed to be here." The guard used a key as big as his finger to snap the lock closed. "Stick your hands out troo da bars."

This last was directed at Howell. His knowledge of Italian had proven valuable on any number of occasions during this trip, although he was furious at being forced to use it here. He pushed his hands out through the nearest pair of bars and held them out as far as he could.

The guard used his knife to cut off Howell's bonds, and to nick his wrists in several places in the process. "Enjoy." He spat upon Howell, and walked away and up the stairs.

"Greetings, friend."

Howell was not feeling the least bit sociable. He was seething. That sodding, two-faced, lying maggot! Howell would kill him. He would rip out the man's throat as soon as they met again.

"My name is Giovanni."

Horace had pretended to be Howell's friend, Howell's comrade in arms, and all the time he had been planning to use him.. Sir fucking Whore's Ass was going to try to kill the Duke, and he would fit up Howell with the blame. He had been manipulated like a puppet. That bastard Horace had pulled the strings, and Howell had danced. Howell grit his teeth, squeezed his fists as tightly as he could, and vibrated with rage.

"You seem bothered about something, friend."

He would trample him into a swine's wallow, and watch that prick drown. He had rip off that fuck's arm and beat him to death with it. He had… Howell suddenly noticed the man opposite him gazing at him with a benign smile.

"Good morning, Signore."

Howell was at a loss. "What in Hell are you so happy about?"

His companion was shorter than Howell, but was even more powerfully built. He continued speaking with his pleasant smile. "Why not? How is it going to help if I feel all sad and weepy?"

Somehow this exchange did not make any difference in Howell's mood. He looked around the dungeon in which he found himself. Each cell had three sides made of bars that had been roughly beaten into shape, probably by an apprentice learning his trade. The fourth was the stone back wall. The door in front of him had a lock the size of his face. The stones behind him were as big as his arse. It was cold down here. He had his doublet, the bloody doublet that Sir Whore's Ass had driven his rapier through, but it did not seem to be enough to keep him warm in this dreary cell.

"It is cold, isn't it?" Howell's companion was even more scantily dressed than Howell was. He wore just a shirt and breeches that only reached to his knees. "What's your name?"

Howell was tired of it all. He had been betrayed, and thrown into this hole. He was furious and cold and now he was starting to get hungry. "Howell Davidson." He went into a corner of his cell and did his best to strip off all the corruption that had been hurled at him while he was in the streets.

"Good day to you Signore Davidson. My name is Giovanni."

Howell gave a sigh. Being in this dreary place was probably better than being hauled before some hanging judge at any rate. "Good morning to you Giovanni."

"If I may ask, Signore, how is it that you have come to visit me?"

"Visit you, eh?"

Giovanni swept his hands to encompass the whole of the dungeon. "Who else owns this place?" e*h*? *eh*? He put his hand to his ear, listening for anyone who might dispute his title. "So, why are you here?"

Howell was starting to warm to this odd character. "I was with a man who claimed to be my friend, and then betrayed me."

"Ah, that must have been a great disappointment."

Howell grunted. Disappointment? That was far, far too mild a word to describe how he felt. It was more like a punch to his gut. As he thought about things in those terms, he began to think about the slice along his side. He stripped off his doublet and immediately felt the cold. He pulled up his shirt and tried to get a good look, but it was at a very awkward angle.

"Here Signore. Turn towards me."

Howell did.

"Ah, that's a cut you be proud of. It is about the length of both your hands, but it doesn't seem to have bled much."

Howell pulled down his shirt to show what a mess it was.

"You misunderstand me, Signore." Giovanni grinned, and they both knew that he was just joking his way out of an error. "What I said was that it was not bleeding much *now*."

*Ah hah* This little exchange helped Howell's mood somewhat, but with all the distractions now behind him, he was keenly aware that his side hurt like Hell. He donned his doublet, but the warmth it provided was nothing like what he was expecting.

"I am from Merico," said Giovanni without preamble.

"Excuse me?"

"It's a small village in those mountains up there." Giovanni gestured up the stairs and towards the peaks that held the city against the sea. "I say small, but that is merely boasting. If you were walking down our path, and sneezed, you would miss it, Signore."

"You know, Giovanni, if you keep being so formal, I am going to have to start calling you Mister… *uh* …what is your last name anyway?"

"Last name, Signore?"

"Yes. Where I come from people have a Christian name, like Howell, and a family name, like Davidson."

"I have never heard of such a thing."

"So you are just called Giovanni, then? Nothing more?"

"Ah, well, they used to call me 'Little Giovanni' sometimes when I was a kid, and 'Big Giovanni' after my cousin was born. We were the only Giovannis in Merico."

"If you are from Merico, then you are Giovanni del Merico," Howell proclaimed, making it sound like a title of nobility.

"Really?" Giovanni seemed rather pleased with the notion.

"If you like." Howell added off hand as he took another peek up the stairs. "Now that you are at liberty, you could call yourself John the Baptist, or King bloody Solomon, and who's to say different?"

Giovanni stood quietly, thinking this over. Then he turned to speak to Howell. "And where do you come from Howell Davidson?"

"London."

"London, Signore? And where would that be?"

It finally struck Howell that Giovanni's rustic accent had to mean that he had not been in Genoa very long, and was likely quite ignorant of the wider world altogether. He had probably broken the peace in some manner, fled from his tiny village, and had come to seek his fortune in the big city. "London is a great seaport on a great island in the Atlantic Ocean." He waved vaguely to the west and north.

Giovanni watched Howell's arm movements with a look of consternation on his face. "Atlantic?"

"Yes, the Atlantic."

"Signore, am I to understand that this Atlantic Ocean is a different sea from the one that I see here?"

"Oh yes, altogether different, and much bigger. Stormier." Howell gazed at the earnest young fellow looking through the bars at him, and tried to imagine just how fast this Giovanni's world was expanding. Ignorant peasant a day or two ago, then the shock of this great city, and now the discovery that there were more seas than the Mediterranean and more ports than Genoa. Giovanni swallowed hard, and scratched at his unshaven face.

This conversation and speculations about his new friend were pleasant enough, but now Howell began to think about the realities of his situation. He knew that he could not stand forever, so he walked over to the stone wall, and sat down. He shifted from one place to another along the wall, trying to find someplace where there was not some nasty point right between his shoulder blades or a void behind the small of his back.

Giovanni sat down against his wall as well. "If you find the right spot, it is like resting on a feather bed, Signore."

"Howell."

"Si. Howell." Giovanni was very happy to address Howell in this familiar manner.

Howell knew that he had have to be content with the least uncomfortable spot on the wall that he could find. The stones were cold. The floor was cold. His back was cold. His arse was cold. His stomach was empty. "Are we fed?"

"The feast comes after the guards' dinner time, when we get whatever juicy bits they did not eat. There are even times when they do not force us to eat anything at all."

They sat there, each lost in his own thoughts. Howell thought about Rosy at the mercy of that monster, Sir Whore's Ass. He could do nothing. It was infuriating.

What little light there had been began to gradually fade away. It seemed that he was doomed to a night without anything to ease his hunger.

Then a different, younger, guard came in and carelessly put a bowl on the floor, about evenly between Howell and Giovanni.

"Thanks," Giovanni said to the guard cheerfully. "This will be good."

The guard rolled his eyes, and shook his head as he left.

Howell and Giovanni looked at one another, but neither made a move towards the food. Both looked at it, though. They were both hungry. Howell knew how long it had been since he had eaten, but he had no idea of how long it had been for his new friend. Probably longer than it had been for him. Howell lay face down on the floor of his cell, and reached out as far as he could. He could get a lot of his hand on it, so he could push it to within easy reach of Giovanni. "You go first." Would he get any of it at all, or would this Giovanni character eat it all?

"Grazie, Signore." Giovanni pulled the bowl to him. It wouldn't fit between the bars, so he put a hand between them to grab some of the food, and put in his mouth. After a few mouthfuls, he pushed the bowl over to Howell. "It is delicious. Try some."

Howell pulled the bowl to him. As far as he could tell, it was still more than half full. He ate in the same manner as Giovanni had. The food was terrible, but welcome all the same. Giovanni seemed completely at his ease while Howell got a few handfuls of the slop into his mouth. Howell licked off his fingers. Then he pushed the bowl back to Giovanni. "You finish it, mate."

Giovanni smiled warmly. "It will be my pleasure."

After they had eaten their dinner, such as it was, the darkness in the dungeon became complete. Howell tried to get some sleep, but the cold floor and the cold walls made it impossible. Then he heard feet on the stairs. Someone was coming down, and carrying a torch.

As the torch got to the bottom of the stairs, Howell saw that it lit three people. One was the older guard, holding the torch, one was the younger guard, and last was a naked prisoner being dragged by the boy. The prisoner seemed have almost no strength in him. He was being pulled by his legs, but he was letting his arms trail behind him without any effort to protect himself. His only animation was turning his head so the stairs hit him on the side of his head and a feeble cry each time one of these collisions happened.

He was dragged to a cage beyond Howell and Giovanni. The younger guard wrapped his arms around the skinny derelict, and held him up. The older guard pulled up the man's limp arms, and wrapped chains around his wrists. When the younger guard released him, the prisoner hung helpless, hard against the bars of his cell that lined the aisle. Howell and Giovanni looked at one another. For once Giovanni did not have a humorous comment to make.

That night the naked prisoner groaned at irregular intervals. Eventually he stopped making any sound at all. Howell did not get any sleep that night.

Some feeble light began to penetrate their prison as the Sun rose. Howell looked around. Giovanni seemed to have slept as well as he could have under the circumstances. The naked man showed no sign of life. All his weight was on the chains around his wrists. His head hung.

Howell should have been feeling sorry for the poor bastard he had heard die in the night, but he was thinking of something else. He was thinking of how he might avoid such a fate himself. "Giovanni, do you want to escape this place?"

"What?" Giovanni seemed shocked. "When I have just starting to get settled in?"

Howell smiled back him.

"All right, all right. If you insist."

"I do insist." Howell started to lay out his plan, but then he heard someone on the stairs. "Grab him when he falls…" was all Howell could say before laying himself curled up next to his cell's door. The older guard came down the stairs.

Giovanni sat in the middle of his cell. "Good morning to you, Signore."

The guard just grunted.

Howell reached out between his bars. "Please, Sir. Please. I am starving."

The jailer kicked Howell's hand, and went to the cell that held the starved man who had arrived last night. The guard poked him, and got no response. He used one of his heavy keys to rap him sharply on the head, and the man never flinched. "Well, well. All done are we?" He opened the cell, and hefted up the prisoner. The chains went slack, and the man's arms fell by his sides, but he showed no sign of life. The guard took off the chains, and dumped the man on the ground. He seized the man's arms, dragged him out into the aisle, and then towards the stairs.

Just as the guard passed, Howell shot out one of his arms, and grabbed the guard's ankle. He put his other arm on the other side of the same bar, and grabbed the same ankle with his other hand as well. The guard dropped the dead man, and turned angrily upon his prisoner. Howell gripped for all he was worth, and stood. The jailer flew across the width of the aisle with such violence that Howell could barely hold on.

The guard's head hit the bars on Giovanni's cell. The jailer slid to the floor. Giovanni shot across the floor of his cell. He grabbed the man's hair, and slammed his head against the floor, over and over, as fast as he could. The first blow had been a surprise, and Giovanni was able to get a good rap, but during Giovanni's next few attempts, the man tensed up making Giovanni's job much harder.

The jailer scrabbled about, trying to free himself, with his head on the floor and one of his legs at the level of Howell's belt. Howell jammed the ankle against the bar with one hand, and pushed down on the man's foot, twisting his leg with all the strength he had.

The jailer was disoriented by attacks from two directions at the same time. Trying to deal with his painful leg, he slacked off the tense muscles in his upper body. Giovanni immediately took advantage, and pulled up the man's head for another great impact on the floor. This time the guard did not recover his strength so quickly, and Giovanni was able to add blow after blow, each of increasing severity. Presently the jailer fell completely slack. Both sides of his head were covered in blood. Giovanni did not stop.

Howell knelt down, and used his grip on the man's leg to pull him closer. He reached out as far as he could, but he could not reach the key ring. He hauled on the man's leg, and pulled him closer. Giovanni finally had to stop his punishment of the guard's head. Howell made another grab, and this time he got his hand on the ring. It was pretty firmly attached, but Howell managed to wrestle it off. He stood.

Giovanni looked up. "I knew I did well when I picked you for my friend."

*Uh hunh* Howell tried several of the keys until he found one that worked. He pushed the dead guard out of the way, and his cell door swung open. He gave the ring to Giovanni. "Get yourself out."

Giovanni went through the keys, looking for the one that would open his cell. "Now what?"

A very good question, that. What was at the top of the stairs? Howell had gotten only the briefest glimpse on his way down, but it seemed that he had been led through the guards' wardroom.. When he and Giovanni went up, they could expect to see any guards that their recent victim had left there. "I am going to get the guard's doublet and hat. You get a length of chain." Howell knelt down, and began to strip the guard of his clothing. "They've seen you for quite some time now, but I am a new face." Howell pointed to their victim. "I am hoping that we will be out the door before they realize that I am not him."

"You are not so well fed." Giovanni smiled at the fat guard.

"It will just have to do."

Howell put on the guard's doublet, and jammed the hat down as far as it would go, angling it so that it would cover his face. Giovanni wrapped one end of the chain around his wrists, and gave the other end to Howell.

"Ready?"

"It is been too long since I have a nice walk in the Sun," said Giovanni.

Howell led off. They went up the stairs, and, sure enough, there was the younger guard lounging at a table.

"Going for a little walk are we?" said the turnkey.

"Oh yes, this should be lovely!" said Giovani in his usual cheerful way.

Then the guard got thoughtful. "I did not hear anything about moving him. Who told you to do it?"

*Ummp Blabla*

Howell's grunt did not satisfy the man. "Wait! What's going on here?" He stepped up to Giovanni and Howell.

Howell turned and gave the man a full-armed blow to his jaw.

Giovanni used his bit of chain as a whip, and rattled the man's brains. The guard flew across the room.

"Run!" Howell led and Giovanni followed close behind as they started across the courtyard. They aimed directly for the guards at the gate.

"Escape!" Howell yelled. He ran right up to one of the guards, and pointed behind him at the entrance to the dungeon. "They are getting away! For God's sake, help the jailers. Do not let them get away!"

The two guards hesitated. Then the young guard that Howell and Giovanni had beaten came staggering out of the dungeon. He was crouched in pain and his head was bloody. He pointed at Howell and Giovanni. This meant he was also pointing at both guards at the gate. "Escape!" He cried.

"Help him. Help him." Howell pushed one of the guards towards the dungeon, and Giovanni followed suit. As soon as these men were on their way, Howell and Giovanni bolted in the other direction.

Their ruse was not going to stand investigation for long, so they needed to get out of sight as quickly as they could. Howell stripped off the guard's doublet, as this would be something easily recognized. Hurrying as they were down the street, it would be difficult to avoid detection. "This way," Howell ordered and led Giovanni into the first alley they chanced upon.

There was no exit. The far end was packed tight with barrels. It might be possible to climb over them, but that would take time. Howell could hear the sounds of the pursuit approaching. Was there any other place to hide? The walls on both sides of them were smooth stone offering no hope of refuge. Then Howell found a door, but his initial elation at seeing a place of escape was just as quickly dashed. The door was locked. The recess in which it was placed was no more than a hand's length deep, and would not cover so much as half of them. Those looking for them would doubtless search each alley as they came to it, so the longer they waited, the more likely it was that someone would find them. "Out. We must get out while there's time."

Three steps from the alley's exit though, Howell froze in his tracks. Several men were talking as they came up the street, and they were speaking in English. Sir Horace had joined in the hunt! Howell instantly reversed course, and went pounding back along the alley directly towards Giovanni. There was nothing for it then, they would have to keep running and hope they could find a way through the barrels at the end of the alley.

Giovanni saw Howell running back towards him, and must have understood the situation, as he turned towards the barrels, and took a few steps. Then he stopped, blocking the alley opposite the doorway that Howell had recently considered as a hiding place. Giovanni smiled.

Howell darted to one side of the big Italian, and urged him towards the barrels, but the big oaf put his shoulder into him, driving Howell into the doorway. Pinning Howell with his weight into what little cover the doorway provided, Giovanni assumed a relaxed pose, and completed his new friend's concealment. He began to chew a fingernail.

"You there!" From the intersection of the alley and the street, a man yelled in English. Howell could swear it was Sir Whore's Ass himself. "Have you seen a fugitive here? A big man running?"

"Scusilo, Signore?" The Italian looked up towards the man questioning him. He did not have to pretend that he was unable to understand English. He spat the chunk of fingernail he had gnawed off into the dirt before him.

"Damned foreigners," muttered the leader of Howell's pursuers. There was the slapping of feet on cobbles and then those who sought him were gone.

Giovanni strolled to the end of the alley to peek at the receding danger.

"How many do you see?" Howell wanted to know.

"Three of them. One with hair like straw."

Then it *was* Sir Whore's Ass that Howell had heard.

Giovanni returned with a reassuring smile on his face. "Once again, Signore Howell, you and I are alone in this great city."

Howell stepped from his shallow hiding place. "Grazie molto, amico." He offered his hand. "You have done me a great service, friend."

The stocky Italian took the tall Englishman's hand, and then enveloped Howell in an enthusiastic hug. When he released his crushing embrace, he slapped Howell's shoulder enthusiastically. "We will do well together, Howell Davidson."

"Amen." One could do much worse in a traveling companion than one who was strong and quick-witted. "You saw that the Englishmen would not know you, but they would know me, so you hid me."

Giovanni smiled. "Something like that."

"Well, we're out of the prison, but still caught in the city. We would do well to make haste." Unlike the pursuit that had gotten him thrown into the dungeon, the men who were searching for him now could not know exactly where he would go. They would have patrols everywhere, though, so he and Giovanni might encounter enemies anywhere they went.

They both looked out tentatively from the end of the alley. They were confronted with a full-sized street jammed with wagons and hawkers. They could see the masts in the harbor.

Howell looked back towards the port. "Perhaps we should just return to the docks, and sign on to a ship bound for America or the Indies or somewhere."

"I have no knowledge of ship work."

"Nor do I." Nor did Howell fancy spending the next year or two having that education beaten into him. Not to mention the poor pay and the terrible food and the threat of disease or starvation or drowning in shark-infested waters after the ship was driven onto the jagged rocks of some barren shore in a shrieking, pounding thunderstorm. There had to be another way.

If downhill wouldn't answer the need, then they would have to keep going uphill until they eventually found a way out of the city. They emerged from the alley and started up the street. They passed a warehouse and then a stable. They passed vendors behind their carts, who tried to tempt him to buy their wares. They shared the street with pedestrians of every kind. These people were the same crowd who had watched the parade, and now thought he was an assassin. None of them showed any recognition, however. Some of the men even tipped their hats to him before catching his odor. He hurried on. He must escape. He must get to Rosy's side.

They came upon a chandler's shop overflowing with all the strange goods that ships seemed to require. A man was outside the front door of this business, sweeping the street. Howell walked past him, and then turned back as though to talk with him. He was really looking to see if anyone was behind him. There they were. Civil guards holding their short spears and all decked out in their flashy uniforms. These unmistakable clothes were meant to impress the people on a festive occasion, not to search for a man who had tried to kill their Doge. Despite how they were dressed, they had a job to do, and they were looking about themselves, trying to find Howell's trail once again.

"Can I help you, Sir?" The man with the broom sniffed dubiously, but then, who was he to refuse a sale?

"Yes. Yes you can. I'd like to see how your stock compares with what I have seen elsewhere." Of course, it was a bluff. Howell had only entered a chandler's shop a few times, and that was in London. Giovanni followed him in.

It was cooler inside the shop, out of the Sun, and that was welcome in itself. It was well lit by the large windows that lined the wall adjacent to the street. The smells that Howell could detect over his own stink were very different than anything that Howell had ever encountered before. Looking at all the nautical gear, one expected to smell the sea, yet that smell was completely absent. He smelt rope and grease and tar. Open barrels offered salt and hardtack bread.

Giovanni gawked. The room was packed tightly with every manner of goods meant to get a ship to sea. Pants and shirts and doublets and even cloaks hung here and there around the room. There were shoes. There were hats and some swords that were fitting for officers. There were knives of many shapes and sizes. There was twine. There were blocks with one, two, three, and even four sheaves for lifting cargo and for lifting the heavy yards holding even heavier sails. There were fixed blocks for the standing rigging that had multiple eyes. There were marlin spikes needed to lace the heavy rope that would run through these blocks. Howell glanced over his shoulder. No one was visible through the window.

"These are only samples, of course." The chandler handed a piece of sail cloth to Howell. "Feel the weight of that, Signore. I can get you as much as you want, cut to whatever size you need. They will hem it at no extra charge." He stood looking eagerly at Howell.

Howell reversed places with the man, as though to get better light on the sail cloth. He rubbed the fabric between his fingers. *hmm*

Several men hurried by in their bright multi-colored uniforms.

Howell walked to the door. "Getting a bit stuffy in here, do not think?"

"Let me open it for you."

"Do not trouble yourself." Giovanni smiled. "I am happy to do it."

Giovanni pulled open the door, and stuck out his head. He took a deep breath to make his excuse for opening the door more believable. He looked both ways to see if the men who had been chasing them were still about. He looked up and said, "Ah, how lovely and clear it is."

Howell immediately understood the meaning of his friend's comment. "Thank you for your time, Sir," Howell said to the Chandler. "I will consider what I have seen here, and then make a decision." He had already decided to get out into the street, and get to a place of safety.

They moved as quickly as they could without drawing attention to themselves until they came to a square. Howell stepped to one side to get out of the traffic, and to consider their course. As much as he wanted to run and force his way through the press, he held himself in check. It was bad enough that he towered over every person that they passed, and that Giovanni was better muscled than anyone they had met. They did not need to draw further attention to themselves by running.

"Gesu Cristo!" Giovanni crossed himself. "I never dreamt there were this many people in the whole of the world, never mind in this one city."

Howell smiled. "So much the better for us."

Giovanni looked puzzled.

"The more people there are, the easier it will be for us to become lost amongst them."

Giovanni laughed. "Well I for one am having no difficulty in getting lost. How is it that you know which way to go?"

"The traffic will be thickest going to and from the city's gates." Howell pointed at the two exits from the square that were across from them. The one on the right had a fair amount of traffic, but the one on the left was positively mobbed. "That one, I would say."

"And I see nothing of the city guards."

"Nor I anything of the cutthroats recruited by Sir Horace."

"Let us be off then." Giovanni plunged into the stream of humanity.

The two men crossed the square, and joined the procession going towards the gate. Everywhere the two- and three-storied whitewashed stone buildings crowded in upon them. Crowded too were the smells of unfamiliar cooking and the all too familiar smells of sweating men and animals, and garbage rotting in the gutters. The air was filled with the shouts of those trying to move their goods, or themselves, and not making satisfactory progress.

They continued in this way until at last Howell could see one of the main gates beyond the thick stream of people before them. Had the city watchmen been put on notice to look for them, or had he and Giovanni been quick enough to outrun the news?

Howell abruptly darted to the side, and ducked behind a stack of barrels. Ahead there were not just civil guards at the gate, but that false Sir Horace himself as well, along with the two guards from the ship that he kept on call for just such a situation as this. Howell needed to come up with a plan of action.

Giovanni wandered casually into his new friend's hiding place. "Well done, Howell. Well done. You are doing a wonderful job of drawing attention to yourself." He became serious. "Everyone here moves at an easy pace, and you will not be noticed if you do the same."

"I beg your pardon," Howell whispered angrily. "The dandy up there is the bastard that betrayed me, and would like to see me hang."

"No shitte?" Giovanni stretched his neck to get a better look over the barrels at Sir Horace and his goons.

Giovanni slumped down next to Howell, and began to gnaw at his fingernail again. "That yellow hair is like nothing that I have ever seen before. Is he deformed?"

"Only in his soul." Howell wiped the sweat off his brow. "That cur took me and Rosy…"

"Rosy?"

"…yes, yes. Not now. Sir Whore's Ass took me to watch the parade of the Doge and his guest, Duke something or other."

"I have heard of this Doge fellow."

"Fine." Howell recreated the event in his mind. "Somebody stuck a pistol out of the crowd next to me and fired it. People thought I'd fired."

"So they started chasing you."

"I ran right up to that trull who's standing at the gate even as I speak. I thought he would help me, but instead he tried to kill me, and then he turned me over to the city's guards." Howell looked again through his crack between the barrels. "It must have been Horace's man who fired the pistol."

*Uh hunh* Giovanni could know nothing of Howell's enemy who now guarded the gate, but he did understand one aspect of the situation far better than Howell. "So you have the vendetta with this man."

Howell looked at Giovanni. "Ven- what?"

"Vendetta. You have sworn to take his life, no? You cannot rest until he is dead."

"Yes." Howell looked between the barrels at the object of his blood feud. He savored the word he had just learnt. "I have the 'vendetta' with this treacherous cur."

Sir Horace finished lecturing his thugs, they dispersed looking fierce and alert, and showed their clubs and knives to all who came near. For himself, Sir Horace had a pistol.

"Would you like some bread?" Giovanni broke a long loaf in two, and offered half to Howell.

Howell was surprised, but he accepted his portion without hesitation. "Where did this come from?"

"A boy with a large basket of these loaves was passing by just now." Giovanni smiled broadly. "Somehow I just knew that he wanted to share."

Howell was less confident than Giovanni. He had nicked more than a few buns during his urchin days on London's streets, and had rarely gotten away without a spat. He hastened to destroy the evidence.

"So, what is your plan, Signore?" Giovanni bit off another chunk of the bread.

"I do not have one as yet." They could try another gate, but there wouldn't be many, and by this time all of Genoa's guards would be on the alert.

The gate was at most twenty paces ahead, so Howell could watch what was going on, but he was effectively shielded from view by his barrel wall. Sir Horace and his crew tried to examine every person and vehicle that left the city, but the local watchmen did everything they could to maintain the commercial flow that nourished this great port. Once a party of guardsmen was actually detailed to seize the wheels of a particularly heavy wagon, and help to get it over the sill of the great gate.

"I wonder what that means." Giovanni pointed at a broadside pasted onto the wall opposite where they were standing. It was decorated with a crude woodcut of a man holding a sword.

It seemed to Howell that his new companion had a unique talent for wrenching his thoughts out of their chosen path and into an entirely new and unexpected direction. "That poster over there?"

"Poster? Is that what it is called? Wouldn't it be nice to know what it says?"

Howell could not read it all from across the busy street, but he picked out enough to be able to get the gist. "It seems that men are wanted to join a regiment of soldiers that is recruiting in Verona. Any man in good health is welcome. They promise to pay well."

Giovanni's mouth sagged open. "You, you, you can read, Signore?"

"Yes, and write, too."

"You have the appearance of an ordinary man, I mean, I beg your pardon Signore, but…"

Howell raised his hands to acknowledge his poor attire and mugged in ironic agreement.

"…and yet, well, you have a skill that even our parish priest lacks."

"Appearances can be deceiving, Giovanni."

"It would seem so. I look forwards to learning what other mysteries you may reveal, Howell Davidson." Giovanni now fell silent. The traffic passed nosily by them. Giovanni seemed to be taken up with his thoughts. "You have said that I could be Solomon as well as Giovanni. I could be dead or alive."

"Come again?"

Giovanni looked about him at the cramped buildings and the throng of people around him. "It would make no difference to this big, big world in which I find myself if I am Solomon or Giovanni, or if I am dead or if I am alive."

Howell could see in this bewildered Giovanni the lonely boy that he had once been dodging along London's alleys. Howell'd had the incredible luck of getting picked off the street by Mr. Riss to be his menial, but until that moment Howell had survived only because he refused to fail. Did this Giovanni have that sort of mettle? "No one is going to care about you, Giovanni, unless you force them to. Make Giovanni del Merico a name to be respected."

"Soldiers are respected." Giovanni nodded towards the poster. "They are great heroes, ravishers, and rescuers, are they not?" He gave Howell a broad wink and a gentle elbow in his ribs. "That sounds good, does it not? Perhaps we could become soldiers, eh?"

"Perhaps."

Giovanni rubbed his hands together in anticipation, but then a cloud passed across his face. "We won't have to fight dragons, though, will we? I want no part of that."

"No, no dragons."

"You smile, Signore, but I have heard the stories."

"Believe me, Giovanni, they are no more than stories. We will not have to fight dragons." Howell was glad that his words reassured his friend, but his attention was really on two skinny, barefooted men who were struggling up the cobblestones towards the gate with a two-wheeled hand cart. One pushed from the rear, and the other hauled on a rope tied to the front. Their vehicle was piled high with manure and rotting vegetables. They probably scratched out a precarious living taking this waste from the city, and selling it to the local peasants as fertilizer.

Howell glanced at Sir Whore's Ass, the object of his vendetta, and made his decision. "Giovanni, we're going to help those two shitte haulers push their cart through the gate." He kicked off his shoes. There was precious little he could do to disguise himself, but bare-chested and with his shirt tied around his long, brown hair as a sort of turban, he made the attempt. At least he smelled the part. "If I am recognized, you make a break." He smiled grimly at Giovanni. "And I will see if I cannot kill that smug bastard before they take me down."

"Very good, Signore, excuse me, Howell."

Howell strode out of his hiding place and slipped beside the peasant who was pushing the cart. "Let me help you, friend." He put both of his hands onto the vehicle, hid his face between his arms, and started pushing. Giovanni seized the rope tied to the front, and began pulling. With their added effort, the cart *clunk, clunk, clunked* over the cobblestones towards the gate and freedom.

"Arresto!" called out someone with a distinctly English accent. "Arresto, damn your eyes! Stop-o!"

The wagon ahead of them came to a halt for examination. The two crap merchants, Howell, and Giovanni were forced to stop, as well. For better concealment, Howell sat on the pavement, but he stole a peek at Horace's toughs. Howell had fought alongside these men against the pirates. He had thought them to be his mates. The next time he saw them, they were searching for him, and Giovanni had been forced to hide him by pushing him into a doorway. Now he knew that they would do Sir Horace's bidding no matter what that meant for Howell.

Sir Horace certainly could not be bothered to do the inspections himself of course, but beyond the legs of his henchmen, Howell saw their leader striding towards him. Howell ducked his head back beneath the cart and hoped he had not been recognized.

Giovanni was still on his feet. Now he left the cart and walked directly up to Sir Horace. "Buon giorno, Signore!" Giovanni welcomed him with excessive enthusiasm.

Dubiously, the gentleman eyed this strange peasant who greeted him. Horace answered nothing. It was plain to all that Sir Horace did not have a single word of Italian.

Giovanni continued earnestly. "I have heard it said that there is a man in town with hair the color of spoilt straw whose nose is bigger than his manhood. Would this be you, Signore?" Sir Horace had no reply, but two of the civil guards stifled a laugh.

Howell kept his head down but watched the proceedings out of the corner of his eye. Giovanni was doing a first-rate job of drawing all of the attention to himself, and thus protecting him. Howell was pleased to discover that the name Giovanni del Merico was one that commanded his respect.

One of Horace's enforcers was on inspection duty before the gate, and now he droned, "Aw right, aw right, aw right. Next!"

Howell's two dung dealing acquaintances bent to the task of moving their cart forwards. Howell stood, and helped. He would be a free man or a dead man in the next few moments.

Giovanni continued to address Sir Horace in a happy tone. "I ask you this so that I may judge the size of the knife that I will need to remove your ability to make babies."

The sniggering of the Italian civil guards was unmistakable now, and Sir Horace could be in no doubt that he was being made the butt of some joke. He stared angrily at Giovanni. "Be gone, jester!"

"Forgive me, Signore." Giovanni spoke as though he was surprised that anyone could have taken offense at his words. "I do not understand you. Do you mean to say that my task is unnecessary, that you have already had your balls cut off?"

The civil guards burst into unrestrained laughter.

"Here!" Sir Horace summoned his arm-twisters to see off this insolent native. Several of the Italian guards sauntered in around them to watch the show.

Howell was left without inspectors just as the shitte-laden cart bumped to a stop on the sill of Genoa's main gate. All he had to do now was add his strength to that of the two turd tycoons, the cart would start moving again, and Howell would be out of the city. He immediately discarded the notion of taking to his heels, and leaving Giovanni to his fate. He had found a true friend, and he was not about to desert him now. Sir Horace's blackguards shouted angrily at Giovanni, and they crowded around him, anxious to use their clubs.

How was he going to free Giovanni? Howell weighed his options. Neither he nor Giovanni had a weapon, so running in and helping Giovanni to fight off Horace's hirelings would mean certain defeat. There was nothing for it, then. He would have to create a distraction which would give Giovanni enough time to slip away. Howell stood, and seized the most solid-looking object in the shitte salesmen's inventory, and threw it directly at his enemy's head. "Sir Whore's Ass, you pompous prick!" All eyes turned towards Howell. "You false bastard! I will see you dead!"

Giovanni bolted. Neither of Sir Horace's brutes could touch him before he was through the line of Italian civil guards, and sprinting towards the gate. Sir Horace cocked his pistol. Horace's thugs scrambled to make their way between the Italian guards without starting a fight with them. Giovanni dodged around one end of the cart, and Howell slipped around the other. As though reading one another's minds, the new comrades paused just long enough to tip the cart onto its side, blocking the gate with its entire slimy, stinking load.

A pistol ball hummed by Howell's ear.

He displayed the middle finger of his hand and shouted, "Until next time, Sir Whore's Ass!"

"Follow me, Howell! I know these mountains well!"

# Chapter VIII

## Mansfeld's Regiment

Howell had been in several large urban areas, so Verona was not a surprise for him, but Giovanni was still trying to come to grips with the whole idea of cities. He had been amazed by Genoa and now he was walking through another. He was overwhelmed by the thought that there might be more cities, many, many more. The crush of people and animals and wagons in the damp, narrow, and echoing streets bewitched the farm boy. They saw push carts filled with exotic fruits or vegetables, and tables filled with meat or clothes or decorations for one's home.

An hour or more passed in their unsuccessful search for the Mercenary Regiment before Howell spotted a very fat, rather well-dressed man with a sword at his waist. He was accompanied by several men each with a long, thick stick upon his shoulder. "Look! Those soldiers are carrying muskets."

"Soldati?" Giovanni took an interest in the group that Howell pointed out. "As far as I know, they could be angels or devils, but as you say, amico."

The musketeers were walking away, and Howell started after them. Giovanni was now on the scent, and no longer distracted by sightseeing. The crowd in the streets parted easily enough for the soldiers, but Howell and Giovanni had difficulty keeping up in their wake.

"Scusilo." Howell shoved people who were closing in behind the soldiers, and blocking his way as he tried to make a path. He was roundly cursed for his troubles. A woman with a large basket spat at him. A man swung a shepherd's crook at his head, but Giovanni deflected the blow.

Howell passed a man in a long cloak, and was suddenly was face to face with the officer who led the musketeers they had been following. He was a beefy man, wearing a black doublet sprinkled overall with shiny silver rivets. He wore a hat with a bright, silver buckle on the front. The baldric suspending his sword was worked through with contrasting thread. Something about his clothes and his presence clashed, as though he was a journeyman brick layer who had stolen his master's attire. All around him stood the men with their businesslike muskets.

Giovanni arrived, and did not realize that Howell had stopped for good reason. He put his shoulder into Howell's back with a cry of "Faster!"

Howell lurched directly into the officer, and all three of them sprawled onto the filthy street. There was a moment of terrible silence as the two lads scrambled to rise, and the officer tried to catch his breath.

"Bastard!" The officer regained his voice, and struggled to regain his feet. "Get this shitte-head off of me!" The soldiers grabbed Howell and Giovanni out of the pile, and held them as prisoners. The officer sat up, and looked at his hands dripping with muck, and his brown-stained elbows. "My God in Heaven!" He glowered at the two friends. Several of his musketeers helped their leader to stand.

He shoved his face into Howell's, and launched into a stream of curses and abuse. He spoke in German, but one could have understood what he meant even without a knowledge of that language. He looked from Howell to Giovanni and back again, pouring out a torrent of profanities the whole time. Howell understood most of the words, but Giovanni could only know that this man was furious beyond imagining.

"Take them!" The officer snapped as he turned on his heel.

The soldiers followed him, forcing Howell and Giovanni to stumble along with them. They were dragged into a small square where a few armed men were standing about. A tavern formed one side of this square, and in front of it were two men sitting at a table covered with a checkered cloth.

The officer snarled something, and went into the tavern.

One of the men at the table came around the end of it, limping ever so slightly. He was above average in height, with long graying hair. There was nothing fancy about his hat, and his jacket was plain, but clean, grey wool. Howell noticed, however, that his right leg was noticeably thinner than his left.

The Sergeant looked over the prisoners, and shook his head. "What the Devil is going on here?"

"Sergeant, these two men were following us, and knocked the Captain to the ground."

The guards chose Giovanni, seemingly at random, and shoved him in front of the Sergeant.

"Is this true, son?"

Giovanni attempted an explanation. "You see, Sir…"

"Sergeant."

"…Sergente, we were trying to follow the gentlemen…"

"Captain."

"…Capitano, si, with these soldiers here…" Giovanni swept his hands around himself. "…but the crowd was so thick that we were having trouble keeping up. My amico called out 'please Sir would you wait for us?' over and over." Giovanni picked out one of the men who was guarding him. "Isn't that right?"

The musketeers just stared at him.

"Go on," said the Sergeant with some irritation.

"Then the Capitano and his men stopped." Giovanni paused, as though savoring the recollection, yet knowing that if he laughed now, it would not be well received. "My amico Howell stopped but I did not see it, so I pushed into him and we both fell onto the Capitano." His face was calm, but Howell knew that he wanted desperately to add a cutting jest.

"Why were you following the Captain in the first place?"

Howell stepped up beside Giovanni. "We were following him so that we could join a Regiment we saw on a poster in Genoa."

Everyone in the square, excepting Howell and Giovanni, burst into laughter at this. Giovanni smiled. Howell frowned. The Sergeant was convulsed. Presently, he got his breath, and wiped his eyes. "Well, that's just perfect. You attacked our Captain so you could join his Company in Mansfeld's Regiment." He chuckled again. He looked at Giovanni and spoke to him in German-accented Italian. "Do you want to join Mansfeld's Regiment, son?"

Giovanni looked at Howell with an expression that said 'why not?' and Howell nodded back at him. Giovanni took the plunge. "Si, Signore, I would be honored to be a member of your Regiment."

"Then come with me." The Sergeant started to limp to his seat behind the table when he saw his pudgy Captain step into the doorway of the tavern. The Sergeant stopped, stiffened up, and touched the brim of his hat. "Sir."

Howell managed to stop in time to avoid piling into the Sergeant from behind, but Giovanni was distracted, and stumbled into Howell.

The Captain was in his shirt sleeves, and was drying his hands on a crisp linen towel. His piggy eyes viewed the scene in confusion. "What do you think You are doing, Blutworth?"

The Sergeant calmly replied, "Sir, I was about to enroll these men in your Company, Sir."

"Enroll them? Why would I want these sons of bitches in my Company, Blutworth?"

Howell took exception to this description of Giovanni and himself. This man might be a Captain and all, but he was certainly no gentleman.

"Begging the Captain's pardon…" Sergeant Blutworth again touched the brim of his hat. "…but you have ordered me to sign what recruits I can find while we are still in Verona."

"Yes, yes, but these?"

"They undoubtedly owe you compensation for the soiling of your clothes, Sir, but, from the look of them, I do not believe that they have enough money." He looked at Howell and Giovanni more closely. "They probably do not have any money at all."

The Company's owner suddenly looked interested.

"With them enrolled, Sir, you could stop their pay as required."

The Captain continued drying his hands as he thought for just a moment. "Very well, Blutworth. Enroll them. As musketeers."

An odd look passed across the Sergeant's face. "Musketeers, yes Sir." Blutworth paused for a moment while he thought about this.

Howell noted the Sergeant's hesitation, and was not encouraged by it.

"Put them with Vlok." The Captain said this odd name in a way that Howell found unsettling. The Captain directed a little smile at a man who was leaning against the wall of the inn, then turned, and disappeared back into the tavern.

Sergeant Blutworth thought about this for a moment before he finished his little trip around the table, and seated himself with his back against the tavern wall. To one side lay a ledger book, exactly aligned with the squares of the checkered tablecloth, with ink and quill neatly occupying adjacent squares. Next to these was a leather sack big enough that it would be impossible to conceal inside both of one's hands. Seated behind these implements was a second man, a gangly fellow, with a thrusting jaw and spectacles perched upon his nose.

"Mr. Bell, let us begin."

The clerk slid the book in front of him so that it remained perfectly aligned with the grid of the tablecloth, opened it, and began expertly working his way through the pages.

"Right then." Blutworth looked at Howell and Giovanni. "Mr. Bell will explain the terms of your enlistment, and if you agree, you will be added to the strength of the Company."

Howell was fascinated with what lay before him. He was going to be a proper soldier. Not the member of some pack of urchins on the street. Not fighting just because he happened to be on a ship that was under pirate attack. He was joining thousands of men who would have a single objective. It was awe inspiring.

Giovanni, on the other hand, really did not know what all this meant, but still smiled at the Sergeant and the man with the book at the table.

The clerk beside Blutworth was ready with the correct page open and his pen in his hand. The Sergeant beckoned his new recruits towards him. "You two come over here."

Before Blutworth had even finished the sentence, Giovanni ambled across the cobblestones, and stood in front of the table. Blutworth's hand shot out to cover the ink pot. He did not fancy a lap full of ink, nor the muster book blotted by this clumsy pair. He looked up dubiously as Howell came up beside his friend, but the Sergeant seemed satisfied that no one was going to topple into him, so he resumed his business.

Howell was ready to proceed, but Giovanni scratched his head. Howell had helped Mr. Riss count out piles of shillings and pence, francs, marks, and kreuzers, but Howell was confident that Giovanni was still anticipating the day when he would actually hold a coin.

The Sergeant thrust his hand into the leather bag sitting on the table between himself and Mr. Bell, and removed a fist bulging with silver discs. He had seized more than his hand could contain, so a glittering shower of leaked kreuzers clanked back into the bag, and clunked haphazardly onto the checkered tablecloth. Using his odd sounding Italian he said, "You will receive 525 kreuzers per month."

Giovanni looked to Howell to see if this was good or bad.

Howell hugged Giovanni like a brother. "Bloody Hell! Five hundred!" From hounded fugitive and ignorant peasant to this!

Sergeant Blutworth smiled wryly and let the excitement subside. "There are also stoppages. You will be fed, but the cost will be deducted from your pay."

"That seems fair," the friends agreed.

"Equipment, certain fees, that sort of thing."

"Yes, Sergeant."

"Si, Serg..." Giovanni was still having trouble with his pronunciation.

Mr. Bell seemed to be bored with all this. "Can we get on with it?" Doubtless he had seen ignorant farm boys react in this fashion many, many times. He dipped his pen into the ink well, and looked at his new recruits. "Name?"

Giovanni stepped forward without hesitation. "I have a name," he said cheerfully. "Giovanni!"

The clerk wrote the name in the book, and held the quill poised over the page. "Giovanni…" he prompted.

Howell grinned at his friend and poked him gently with his elbow. "Go on. Go on."

Giovanni grinned. "Giovanni del Merico!" he proclaimed proudly.

"…del Merico." Mr. Bell mouthed the name as he wrote. The lanky clerk now whipped the book through a half turn and pointed to a spot next to some wet ink. "Put your mark there." This snippet of Italian was oddly-accented, too, but it was not with German. Howell thought he detected a Scottish inflection.

"My mark, eh? Certainly. Si. My mark." The phrase 'Put your mark there' obviously meant nothing to him. Howell gently eased Giovanni to one side and took the offered pen.

Before Howell could write, though, Sergeant Blutworth blocked the page with his hand. He looked at Giovanni. "Are you content with this, son?"

"Content with five hundred of those round shiny things? Si, I think I can manage."

"Very well, then." Blutworth seemed a bit irritated with Giovanni's joking style of speech, but took his hand off the page.

Howell leaned over, and wrote 'GdM' next to Giovanni's name.

Now Mr. Bell did something that lit up Giovanni's face like sunrise on the first day of Summer. With a great sigh of relief, the clerk raked up the disorderly, spilt coins, and began to place them, very precisely, five at a time, into the squares of the cloth directly in front of Giovanni. He filled the five squares closest to Giovanni and then started on a second row. He repeatedly plunged his hand into the sack to retrieve more coins until he had completed five rows and five columns with five coins in each square. Then he carefully replaced the left-overs in the leather bag.

Sergeant Blutworth held his open hand over the coins, looked directly at Giovanni, and said "Accept this signing bonus and our business is done." He waited.

Howell started to say something, but it was unnecessary.

Giovanni removed his hat, leaned down, and carefully herded his first silver coins off the table and into the crown.

Mr. Bell now took the pen from a grinning Howell, and turned the book back to where he could write in it. "Name?"

"Howell Davidson." The clerk slowly repeated the name as he wrote. "An Englishman in the Company, eh?" His Scotch accent was unmistakable now. He reversed the book, and offered the pen. "Your mark, if you please."

Howell took the pen confidently, and signed his name. He handed back the quill. "I can read and write, and do sums, and speak German and French as well as Italian."

"Can you, now?" Mr. Bell seemed genuinely interested. He thrust his hand into the sack and began his careful process of placing one hundred and twenty-five silver kreuzers expertly before Howell. "Perhaps there will come a time when I could use talents such as yours."

Howell grinned. It never hurt to make a friend of the man who paid out the money. When the scribe signaled that each of the twenty-five stacks was correct, Howell swept his coins off their squares. Howell was in a position where he could buy his way into the depths of drunkenness and sexual oblivion if he wished. While this was very tempting, and would have been at the top of his list at one time, now he had a different mission.

Sergeant Blutworth looked at Howell and Giovanni. He seemed very earnest, even though he had probably given this advice many times before. "Some men drink up all their pay, and some blow it all on the strumpets, but if you manage your money, you'll get by."

This is what Howell needed to do. Now he was not just an individual, making choices only for himself. Now he wanted to see Rosy again, and he wanted to see Sir Whore's Ass dead. This would likely take all the money that he could gather.

"Right!" Blutworth examined his new recruits. "You two will be going with Corporal Vlok."

"Excuse me, Sergeant?" Howell was unsure of what he had said. "Corporal Fuck?"

"Vlok. Pronounce the 'L', Englander." The way that Blutworth said it, the name sounded somewhere between 'Fluk' and 'Flawk'.

"Corporal Vlok, yes Sergeant."

Sergeant Blutworth motioned to a hulking lout with a goodly sized stick in his hand. "Vlok, take these men to the barn. I will be along presently."

"Yes, Sergeant," Vlok answered reluctantly. He did not really touch the brim of his hat, but just sort of waved his hand in the vicinity. "You two follow me." He did not wait for an acknowledgment from his two fresh faces, but just led Howell and Giovanni from the front of the tavern, and down one of the streets that exited the square. "You think you are big heroes now, don't you?" Vlok did not even bother to look over his shoulder. "You think they put your name in a book, and you are soldiers." He whipped around to confront them.

Howell and Giovanni had to stop to avoid crashing into him.

"You've got to learn how to use a musket, and, as like as not, you'll shoot yourself before you are done. If you manage that all in one piece, you'll have to carry the bloody thing for hour after hour on hot, dusty roads. You'll probably try to run away as soon as you hear the enemy firing, and that will give me a perfect excuse to kill you. If you are stupid enough to actually make it into a battle, you'll shitte your breeches when you see it up close. Men yelling. Wounded screaming. Smoke and blood everywhere. You turds will never make it." He turned without another word and led them down a side street. *pfft*

Howell's reaction came from deep within his soul. Battle. He was going to be a part of a real battle.

As the three men progressed, shouting erupted from a building in front of them. There were half dozen men just inside the door, sitting on the floor around a blanket. On the blanket were scattered silver and copper coins, and two small, white cubes. "Corporal Vlok!" called out the first of the gamblers to see them approach. Howell and Giovanni entered what was probably a warehouse most of the time, but now housed more than a dozen likely lads about the same age as Howell and Giovanni. Everyone in the room got to his feet when Vlok came in. Some of them looked glad to see him, some looked wary, and some looked positively fearful.

"So, what are you dogs doing?" He kicked the blanket that held the coins and the dice into a crumpled heap. "Just fucking off, aren't you?" He walked around the space, looking daggers at everyone. "You'll do what I say, exactly what I say, or you'll be a dead man." He moved from man to man around the circle that had formed inside the barn. "You'll do what I say in battle, or the enemy will kill you. If you try to run away, then I will kill you." He carried a stick in one hand, and slapped it into his other palm, over and over. "You'll do what I say, or I will cut you up with this."

Now he stopped, and concentrated his attention on those who looked most horrified by him. Vlok jammed his face into that of the youngest and most fearful of the men in the 'barracks'. "You! I am going to do for you." He held his stick just inches from the boy's nose. "I am going to make you curse the day that they wrote your name into that bloody book. I am going to make you curse the day that you were born."

"How about me?"

Vlok whipped around to see one of the new recruits looking down at him.

"Do you think you can dump your bullshitte on me?" asked Howell.

The Corporal was confused. It seemed like he had never confronted a situation like this.

"Well?"

Now Vlok was furious. He raised his stick to give Howell a mighty blow, but Howell's hand shot out, and grabbed his arm. Vlok twisted for a moment, but Howell had his hand locked onto Vlok's wrist, so he could do nothing. He tried to punch Howell with his free hand, but Howell swept out his foot, cutting Vlok's legs out from under him, and sent him crashing to the ground.

Howell still held Vlok's wrist, so he pulled the stick out of his hand. He threw it down hard, aimed at Vlok's head so he had to cross his arms to keep from being hit in the face. "It would seem that you cannot dump your bullshitte on me," said Howell calmly.

Sergeant Blutworth stepped into the barracks, and took in the scene. He stared at Vlok.

Vlok grit his teeth, and got to his feet.

Blutworth continued to stare.

Vlok continued look furiously at Blutworth, but finally touched the brim of his hat. He had been forced to provide the proper respect to a superior officer.

Blutworth pointed out the door. "Report to the Captain." He turned away, and looked back at the men in the room around him.

Vlok stood for a moment, looking with rage at Blutworth's back, and then turned and left.

A noticeable calm filled the room. Everyone seemed to give a great sigh of relief.

Sergeant Blutworth addressed the room. "We're going over the Alps into Germany men, so get ready. Most of you will catch your death of cold dressed like you are now, so visit one of the vendors in the area. You've all gotten your signing bonus, so there should be no trouble with the cost."

"Yes, Sergeant," the soldiers answered as one, but several of the gamblers looked at one another nervously.

"Dismissed."

So Howell was going to cross the Alps. Yet another adventure. It also meant that he was going to get closer to the Countess where he might be able to be reunited with Rosy.

As the soldiers dispersed, Blutworth eyed Howell more carefully than he did any of the others. Perhaps in Howell he thought he saw something out of the ordinary. "So, Englander," began Blutworth conversationally, "You are a long way from home."

It was a perfectly innocent question, just something to pass the time, but Howell realized that an error here might eventually link him to the assassination attempt, and to Sir Horace's effort to blame it on him. "Yes, Sergeant." Howell tried to sound off hand. "And I imagine that I find Italy as interesting as you must. Unless I miss my guess, you are not a local either. How do you come to be here, Sergeant?"

"Some little Italian Duke hired the Regiment last year. Seems this Duke was having a difference of opinion with the King of Spain about a plot of land hereabouts."

"King?" Howell was thrilled by the notion. "Spain?"

"Then the Duke and the King kiss and make up, and suddenly there's no work for honest soldiers like ourselves."

Howell was confused and swept his hand around them, "So what are all these men for?"

"We've been hired to support a rebellion that's broken out in Bohemia. We'd be out of work except that the Emperor is telling all the Protestants in Bohemia to become nice, obedient Catholics, and I mean now. The Emperor had some representatives or viceroys or what-have-you in one of the towers in Prague Castle. The leaders of the Bohemians who've got their noses out of joint marched right up the stairs to the room where the Emperor's men did their work, and threw them straight out the castle window!"

"They what?"

"It is God's truth. They absolutely seized the Emperor's lads, and chucked them out a second floor window. But they did not die."

"They did not?" Howell gasped.

"No," said Blutworth. "Barely hurt. They were spared because they fell into a great heap of crap." He started laughing.

Howell grinned. Taking pleasure like this at the expense of the greatest nobles in the land pleased him enormously. He could not help but try to assert himself. "Saved by shitte, you say?"

"Preserved by poop," added Giovanni.

The Sergeant continued laughing. "Exactly. So the Bohemians cried out for help, and Count Fredrick, who rules over some place or another in Germany, hired our lot to provide it. That's why Colonel von Mansfeld is filling up the ranks of his Regiment again."

Howell was fascinated. Armies recruited in far-off lands. Italian lords at war with the great Kingdom of Spain. Rebellion! This soldiering was beginning to sound positively grand.

Something else seemed grand as well. Howell could smell food and hear the cook and his assistant preparing their evening meal. He and Giovanni had scrounged very little on their walk from Genoa, and a meal would be most welcome.

While they talked and waited for their soup, another man hovered nearby. He looked to be the youngest in the bunch, small, thin, and pimply-faced. He watched them anxiously, but said nothing. Giovanni waved over this undersized comrade of theirs. "He's new like us." Giovanni tried to be as unthreatening as he could, but the boy hung back, his eyes wide.

Howell walked over to him and extended his hand. "Pleased to meet you."

The lad hesitantly reached out, and took Howell's hand. They shook solemnly. The young man responded to Howell with a few words that were completely incomprehensible.

*Uh hunh* Howell turned to Giovanni in confusion.

Giovanni scratched his beard. "I cannot make that out. It doesn't sound like anything I have ever heard."

Howell returned to the youth. He was probably little different in age from Howell, but he was so small that he seemed to be no more than a child. Something else slowly became clear, too. Every drinking, laughing, cursing man in their 'barracks' was noticeably smaller than Howell and Giovanni. Howell now thought he understood Sergeant Blutworth's odd reaction when the Captain had said to sign them as musketeers.

The bulky Corporal Vlok lumbered into the middle of the room, and looked over his command gravely. "I have new men. Where are these puppies?"

"Here, Corporal." Howell answered for himself and his mates. He, Giovanni, and the lad stood together, awaiting whatever might come next. Howell was impassive. Giovanni was eager. The boy might be terrified or mystified, depending on how one interpreted that pimply little face.

"Here Corporal," repeated Vlok in a mocking tone and he scowled, "What the Devil is this?" He shuffled through the straw that covered the floor to his new men, and examined them in disbelief. "A giant clown, a clumsy ox, and a poxy little mouse. I will be damned! Sergeant Blutworth, is this your idea of a joke?"

Blutworth did not answer. He remained in the shadows along the wall.

"Begging the Corporal's…" began Howell.

"Silence, you great buffoon!" Vlok was no taller than the other men in the room, so Howell and Giovanni looked down upon him as he stepped up to them to continue his rant. "If I wanted to hear monkeys talk, I'd ask!" Vlok's character was rather like that of a younger version of their Captain. Not so fat as his commander, but just as insulting. "You will speak when spoken to, and not otherwise! Do I make myself clear?" Vlok slapped his stick into his hand.

"Yes, Corporal," Howell and Giovanni mumbled in unison. Their new friend did not speak.

"What did you say?"

"Yes, Corporal!" they boomed. This time all three of them said the same words.

The confrontation had attracted all the rest of the men of the squad. They were gathered in a circle to see how the Corporal dealt with these raw recruits. Vlok regarded the newcomers severely. "If you are going to work for me, then you need to follow my rules."

"Yes, Corporal," the three boomed again.

"I give the orders and you follow them."

"Yes, Corporal!"

"And nobody holds out on me."

Howell barked, "No, Corporal," but his two companions merely repeated, "Yes, Corporal." It was difficult to be sure what was the correct answer.

"You curs!" Vlok yelled viciously. He punched Howell in the stomach, catching him off guard. When he tried the same thing on Giovanni however, the stocky Italian was tensed up and ready, so Vlok's blow hurt them both about equally. Vlok took a step back, waggling his hand to shake off some of the pain. "Strip!" he ordered. "No one joins Mansfeld's Regiment without proving his courage. Strip I say!"

"Strip! Strip! Strip!" The men in the ring started chanting in a practiced cadence. "Strip! Strip! Strip!"

Vlok stepped forwards to enforce his orders, but did not get far. Howell put his palm on his forehead, and shoved him back.

Giovanni winked at another man who advanced upon him. "Oh, Sweetie Pie! You are going to get such an eyeful." The man recoiled in confusion.

"Strip! Strip! Strip!" Someone had found a tub, and was beating the bottom as an improvised drum. *Thump! Thump! Thump!* "Strip! Strip! Strip!"

Giovanni was content to bow to the inevitable, and began to remove his things, but Howell most emphatically did not. His hands clenched. He eyed Vlok, and weighed his chances. The other men in the room had been with Vlok for some time, and Howell was brand new. It looked likely that they would close ranks around the Corporal in a fight, even if it was only to avoid punishment if they did not. What would happen if he took on these poor odds in a furious rush of fists and knees? The urge to split the lips in those smirking faces before him was almost irresistible. The standoff might have lasted for quite some time, except that Vlok now concentrated his efforts on the poor little newcomer. Unable to understand the orders he had been given, the boy was following Howell's lead, and standing defiantly clothed.

"Strip! Strip! Strip!"

There was nothing for it then. If Howell did not yield, it would go hard with the lad, and that wouldn't do. Howell took off his doublet, and dropped it on the floor.

*Hooray!* The squad cheered. *Hooray!*

Vlok continued his furious look, and his efforts to relieve the pain in his hand.

"Strip! Strip! Strip!" *Thump! Thump! Thump!*

As Howell peeled off his garments, he nodded at the small man.

He realized what must be done. All the recruits were naked in a moment.

"Over there!" Vlok pointed the mass of soldiers towards the open space in the middle of the room. "Form up!"

There was a great whooping and hollering as the men scampered across the warehouse, and into formation. They made two lines, facing each other, and distant from one another about two times a man's full span. There were perhaps a dozen raucous men on each side. Every man in the formation was armed with a belt or a wooden switch or a length of cord held as a whip.

Vlok's pals shoved the three new men to one end of the chute. Howell did not like the look of this at all.

Corporal Vlok appeared at the far end of the thoroughfare. "Soldiers! Comrades! These dogs seek to join our ranks. Are they worthy?"

"Test them!" The men in the lines clamored. "Test them! Test them!"

Suddenly the tumult ceased as Sergeant Blutworth appeared next to the recruits and the men guarding them. He looked at them calmly, and then he looked at the parallel lines of the formation. "No knives," he said without emotion. "Drop them at your feet." As the blades rained down, he began a slow walk between the rows of men, casually examining them. He caused the cook's assistant to cast aside a bit of firewood that he had been holding as a club. Sergeant Blutworth came to the end of the gauntlet, and stepped behind Corporal Vlok. "Proceed."

"Proceed," muttered Vlok, and then he drew a deep breath. He looked around him, resuming command. "Are these sons of bitches worthy!?!"

"Test them! Test them!" the men called.

"Then begin!"

Vlok's henchman clamped down on Howell and Giovanni to keep them from moving as they shoved their little friend into the gauntlet. It was deliberately done with such violence that he immediately slipped, and fell. This put him into a very bad position, for the first few men in both columns had a clear shot at the lad while he was on his hands and knees.

"Get up!" yelled Giovanni.

"Get up!" Howell tried to break free from the grip of one of Vlok's ass kissers. "Get up!"

They were both being firmly held, or they would have rushed to the boy's aid. The blows rained down upon him with the sickening snap of whip on flesh. His back sprouted welts and bloody lines as he struggled to regain his feet. Once he was up, though, he got his bearings and suddenly sprinted towards the end. He surprised the crowd with his speed, and got past more than half of them before anyone realized he was so quick. Then someone stuck out his foot, and their friend went flying again. This time, however, he kept his head, quickly leapt to his feet, and finished the course. He was bleeding from innumerable cuts, and he would show a lot of bruises tomorrow, but he had done it! As he fell in behind Vlok, he was grinning.

Now Giovanni was thrust towards the clamoring soldiers. The push was expected, and the Italian farm boy was far too big to be shoved to the ground, so Giovanni did not lose his feet. As he strode into the chute, the men in the ranks were surprised by Giovanni's size. They had never seen a musketeer so big. The instinctive reaction of the men lining the gauntlet was either to hit Giovanni low, where it was easier for them, or to swing for his back, which was at eye level for them. This meant that Giovanni's face was ignored, and he stood up straight to take it further out of reach. While their little friend had needed one hand for his face, and one for his privates, Giovanni could cover his vital area with one hand, and use the other for general fending off. It was not a walk in the Sun for him, though, not by a long shot. His back and legs were swiftly crisscrossed with scarlet lines. Some of the blows he took were heavy enough that he staggered. By the time he got to the end of his ordeal, he was veering back and forth from one side of the gauntlet to the other.

Howell had formulated a course of action while he watched his comrades' ordeals. He watched, and by the time Giovanni had finished, he was ready. His Captain had called Howell a son of bitch, and Vlok was doing the same. You think we're dogs? I will give you dogs! *Oow! Oow! Oow!* Howell's defiant barking filled the building.

He did not wait to be propelled into the maelstrom. The instant that Giovanni passed Corporal Vlok, Howell whipped his elbow back into the face of the guard behind him and took off running. He did not try to stay in the middle, though. What he had seen Giovanni do by accident, he did deliberately. His first move was to crash directly into the line of men on his left. No one on that side could get a clean cut at him, and no one on the other side could reach that far. Naturally, the press of men on his left shoved him away so that they could get a full swipe, but Howell acted as though he was staggering as Giovanni had, and proceeded to plow into the line on the right side. He did not escape without injury, but his progress bouncing from one line to the other meant that almost no one got a full-armed swing. Howell arrived in front of Vlok with only the slightest of injuries.

As Howell popped out of the chaotic gauntlet, he found the Corporal standing before him, waiting, his arm raised. Vlok was holding a belt, with the buckle at the end. Faster than thought, the belt swung. Vlok was aiming for Howell's head, but because he was so much taller than the Corporal, Howell was able to turn away and take the blow high on his back. It was full contact, though, and it unbalanced Howell.

Vlok cocked his arm for another go, but this time Howell was ready. As the belt swung towards him, he raised his arm to deflect it. His only intention had been to block the blow, but the belt wrapped itself around his arm. Howell's reaction was purely instinctive. He jerked the belt out of Vlok's hand, and stood there glaring at the man.

Sergeant Blutworth stepped between them, and put his hand on the belt. He looked back and forth at Howell and Vlok, and gently took the belt out of Howell's hand. Then he turned to the multitude. "They have proven themselves! Greet your new comrades."

The squad whooped and hollered. Giovanni and their friends converged upon Howell, congratulating him and themselves on their survival. Howell relaxed. Between the laughing faces and the heaving bodies, he caught a glimpse of Vlok. The Corporal looked fiercely at Howell.

*Oow! Oow! Oow!* Howell barked directly at Vlok.

Sergeant Blutworth made his way through the crowd. "Well done, one and all." He congratulated both the men who had formed the chute and the new recruits on their success. He ended standing next to Howell, and slapped him on his muscular back. He offered Howell his great paw, and the two shook hands warmly. "Welcome to Mansfeld's Regiment, Davidson."

# Chapter IX

## Becoming a Musketeer

Move it, you lot." Corporal Vlok was none too patient getting his students onto his improvised training ground. On any other day it would be called a farmer's field, but today it would be the place where Howell would learn his new trade. "Move, move, move."

Howell, Giovanni, as well as all the other untrained men that their Captain had managed to collect, walked out onto this large, empty ground with a skirt of woods at the far end. It was blistering hot, with flies buzzing and tiny gnats or somesuch drifting by in clouds. Sweat trickled down Howell's face. He had bought a red doublet for the trip over the Alps. I had seemed very elegant at the time, but now he found that it was much too hot for this weather.

They were all of them carrying a load of mysterious equipment. First there was the musket itself. It was perhaps four feet in length overall, with wood at one end and a metal pipe at the other. Howell had seen enough of firearms to know that the wooden end was how one held the weapon against your shoulder, and the other end was the barrel, where the balls came out. It was all of twelve pounds, perhaps more. There was a little, curving iron bar sticking out from perhaps a third of the way up the musket's length, and this had a pair of tiny parallel plates at the end closest to him. Howell could pull the trigger, making it pivot back towards him. The harder he pulled, the closer the tip got to him. When it got all the way back, it touched a little metal shelf that stuck out of the right side of the musket. It looked rather like a cock pecking at his food.

None of them had a clue as to the purpose of all the rest of what they carried, however. There was a broad leather strap that went over their left shoulders and met on their right hips. At this low point were collected several items. There was a small, leather pouch that was very heavy for its size. There was what looked to be a cow's horn. A bundle of thick, grey cord was shoved through the strap. The strangest part of all this equipment, though, was the fringe of wooden decorations that hung from both the front and the back of the strap, each on a pair of stout strings.

"Right!" Vlok called them to order. "You think you are soldiers, eh? You put your mark in a book, and took some money, so you think You are musketeers. Well, all of you no more than clumsy, thick-headed clods. I will show you what it takes to be shotte."

He held the musket horizontal at his waist. He did something at the middle of the musket. He reached down to his hip, grabbed his cow horn, and pulled a plug out of the narrow end with his teeth. Then he did something with the cow horn near the middle of the musket, put the plug back into the cow horn with his teeth, and let it drop so that it hung suspended from its cords. He did something at the middle of the musket again and blew on it. He let go of the musket with his right hand and let it swing in his left hand. He put the butt on the ground so that the musket stood next to his left leg. He grabbed something else off his right hip, and put it in his mouth. He grabbed one of the wooden decorations, and popped off the top.

Howell realized that this was not a decoration, it was a bottle.

Vlok held the mouth of this bottle at the lip of the barrel, and turned it upside down. He wriggled it, and tapped it a little. Then he just let it drop to dangle on its cords. A stick seemed to appear from nowhere in Vlok's right hand. He spat whatever was in his mouth into the barrel. He shoved the stick down after it into the barrel, and jammed it down a few times.

This time Howell could see that he flipped the stick end-for-end, and then hid it somewhere near the muzzle.

Vlok swung the musket horizontal at his hips. There was a length of the thick, grey cord between the fingers of his left hand that Howell had not noticed before, although it must have been there the whole time that Vlok was loading. Both ends of this cord was smoldering with a dull coal. Vlok took one end, and jammed it into the little jaws of the curving iron 'cock' on the top of the musket that Howell had noted before. Vlok pulled the trigger a few times, watched what happened a few times, and moved the cord a few times. He blew on the end of the cord to get a bright coal. He lifted up the musket, held the wooden stock against his shoulder, and aimed over the heads of his pupils. He did something at the middle of the musket, and grasped the weapon tightly. He eased the trigger back, and the iron 'cock' swung back towards him. Suddenly there was a flash of light and smoke right next to Vlok's right eye, and almost immediately, a terrible *bang!* and a cloud of smoke at the muzzle. The hum of the ball going over their heads was a terrible surprise. Vlok put the butt of his musket on the ground. "Right. Now you do it."

Howell looked down the line of the men who had followed him onto the field. Some stared. Some looked mystified. Some had dropped their jaws in addition to one of these expressions. Most looked at the men on either side of him. No one had the slightest idea of what to do next.

Howell was furious. "We could not possibly follow all that."

"We?" Vlok sneered. "We? Do you speak for everyone now?"

"He speaks for me," said Giovanni.

Howell was not going to let this hulking shitte get the better of him. "We will work this out together. Hold your musket level like this." He hefted his musket up so that it was horizontal at this waist. "What was the next thing you saw? Anyone?"

Vlok grinned.

Giovanni spoke immediately. "He grabbed his cow horn. He pulled out the cork with his teeth."

"Yes." One of the men pantomimed emptying some powder into his pan. "He put the little end over middle of…" *ahh*

"Musket, you fool." Vlok's tone was nothing short of mocking. "It is called a musket, you simpleton, which is why we, or at least me, are called musketeers."

"Yes Corporal. It is a musket." Giovanni's tone was just this side of insolent.

Howell looked down at the area on the musket where the Corporal had been doing whatever was he had been doing. Howell looked at the little shelf coming off the side of the musket, and he saw a tiny cover on top of this was that rotated horizontally. He turned back this cover to reveal that the shelf was actually more like a miniature pan, and on one side of it there was a hole that lead directly into the bottom end of the barrel. "Halloo. Pull back this little hatch here, mates." Howell held up his musket to show them where he meant.

Up and down the line the men saw what Howell was talking about. One of them grabbed his cow horn, pulled the cork with his teeth, and tipped the narrow end into the bowl-shaped projection. A fine, black powder flowed out, and quickly filled it. He tilted up the horn as quickly as he could, but a goodly amount of the powder spilled upon the ground.

*Ha!* Vlok made another of his helpful comments.

"Then he blew it off," one of the men said.

One of the students blew on it, and a cloud of black dust covered his face.

Vlok thought this was hilarious.

"Let's all try the same thing, but close the little hatch before you blow on it." Howell started on the task.

First Giovanni and then the rest of Howell's mates did what Howell had shown them. There was still some powder spilt on the ground, but everyone got something in his pan, closed the lid, and blew it off without a mishap. Very carefully, each man placed his cow horn back on his hip.

"Now what?" Howell could not remember.

A man down the line banged on the side of the musket, presumably to get some of the powder from the pan into the barrel. Howell and the rest did the same.

A man turned his musket vertical, and placed the butt next to his left leg.

Giovanni looked at what he had had done, and piped up. "Si, that's what he did, and then he took something from his right hip, something hanging from the bottom of this, ah, this big leather strap."

"Bandolier, you git."

Giovanni smiled at Vlok. "Is that what it is called? Bandolier? Thank you Corporal. You are most helpful." Giovanni turned to his mates. "Whatever he found, he put in his mouth."

Each man searched his kit. "How about this sack of lead balls? This might be it," said someone.

"Yes," Howell said, "That must be it." He took out one of the balls, tossed it gently into the air, and caught it again. It weighted about an ounce. He popped it into his mouth, and discovered that it had a slightly sweet flavor to it.

Vlok did not look happy.

"Den un…" Howell spat out the ball into his hand. He ran his fingers along the fringe of powder containers on his chest. "Then one of the little bottles here."

"They are called chargers." Vlok did not seem to be taking as much pleasure in correcting his students now.

Howell put the ball back into his mouth, picked one of the bottles, and popped off the top with his thumb. It was filled with the same black powder that he had seen when he had filled his pan on the side of the musket, except that this was coarser. Up and down the line, his friends were watching Howell. He held the edge of the bottle at the edge of the muzzle, and tipped the bottle up. He gave it a shake, and carelessly let it drop. He spat the lead ball into the barrel.

While his mates were following his example with the charger and the musket ball, Howell searched for the stick that Vlok had so suddenly produced. He found it right underneath the barrel, in a slot made just for it. Howell gently pulled this out. He had never done this before, and there was no telling what might be required.

It was longer and longer. Howell had to shift his grip down the stick as he just could not stretch his arm any further. The stick was at least as long as the barrel itself. His mates watched this, fascinated. Finally it came free.

"The stick is right under the barrel here." He held up his musket so all could see. Howell started to put the stick down the barrel.

Giovanni spat his ball into the muzzle. "Howell. He spun his stick to point the other end at the barrel. Why would he do that unless you are meant to use this end to do whatever it is that we're doing?"

*Ahhh* Howell saw that this end had a little copper shoe upon. "You've got it, amico." Howell held up his ramrod and called to his mates, "Spit the ball into the muzzle and then turn your stick end for end." He demonstrated. "Now put it in the barrel and tamp things down." Howell put his stick down the barrel and jammed it a few times. He continued his lecture. "When you are done, pull the stick out of the barrel, turn it 'round, and put it back into the slot."

Up and down the line, the men were doing what Howell had shown. Then one man cried out, "Damn." He was holding half of his ramrod in his hand. He had tried to put it in at an angle, instead of straight down, and the wood just could not take the strain.

This improved Vlok's mood considerably.

Howell looked at Vlok with confidence. He and his mates had sorted what was necessary even with this man's best attempt to embarrass them rather than perform his office and teach them. Howell was sure that this was the first time Vlok's tactic had not worked. He had brow-beaten any number of pupils, calling them fools and idiots and halfwits, and making himself feel important in the process. Now Howell and his mates had beaten him. There was only one step in the process left to complete. "How about some of that burning cord there?" he said to Vlok.

Vlok sneered. "It is called match." He was salvaging what he could from his failure.

"Match then. We will need a bit of that to finish, *eh* Corporal?"

Vlok hesitated. Then he started berating the men again. "Cut off a foot of your match. Hurry now. You know what a foot is, am I right?"

Giovanni looked down at the ground. "Which foot am I supposed to cut off, now?"

The men who were lined up with Giovanni laughed.

"Quiet in the ranks!" Vlok did not understand Giovanni's sense of humor in the least.

The men looked to the length of cord that hung from the low point on their bandoliers. Each of the recruits began to unroll a short length of it.

Vlok walked up and down the line, cursing the men. "I cannot wait all day. Lively! Suppose cavalry was bearing down upon us?"

Finally, each man had some of the match cord between the fingers of his left hand. Vlok went to the man four down from Howell. This man held the end of his match next to the end of Vlok's and blew upon it until it started smoldering.

"Every eighth man." Vlok walked away from Howell. "Light your match from the man four places from you." He continued down the line, starting as many matches as required. When he was done, Vlok prudently stepped behind the rank of trainees.

Howell could see that Vlok was giving the orders, but he was not leading the men. To do that, one needed the respect of your command, and that required effort. Effort other than scolding them because they were still learning. Effort other than trying to make yourself feel important at their expense. It required caring about them, and sharing their trials. Howell felt that he was doing that, and it made him feel a respected part of this group. It reminded him of how he had felt in his urchin days when he had stepped up to a challenge.

"Blow your matches."

The men held up their left hands and blew the ends of their matches' red hot.

"Cock your matches."

Now the men were at a loss. They had never seen what he was talking about, except, perhaps, at lightning speed.

"Those little jaws at the end of the cock you asses."

Howell saw what was meant. At the end of the iron arm, what both he and Vlok called the cock, were those two little parallel plates on the curving arm he had noted earlier. These were just the width of the match, so Howell pressed his in.

"No, you greenhead, the fire goes on this end." Vlok was abusing a trainee further down the line, reversed the man's match in his cock. Several other men had to do the same thing to avoid a tongue-lashing from the Corporal.

Howell pulled the trigger carefully, not sure what to expect. "Now pull the trigger gently to make sure that your match will touch the pan." The cock moved only slightly towards him. He repeated the move, pulling ever harder on the trigger to draw the cock towards him until the match touched the pan. It was a trifle too long, however, so he adjusted it backwards through the jaws. "Move your match in or out so that it hits right in the middle of the cover." He released his trigger, and the spring pushed his cock back up to its vertical position. He looked up to see Vlok standing right in front of him.

"Right clever dick, aren't you? Think you can teach these piglets better than I can, don't you? You think that the Sun shines out of your arse. Well, we will see." He snapped around to address all the men. "Blow your match."

The men bent down to blow on their matches one more time so they glowed.

"Present your piece."

Every one took their aim.

"Open your pan."

Howell swung open the little cover on his pan.

"Give fire!"

Howell pulled his trigger all the way back. There was a blinding flash next to his right eye and acrid smoke in his nostrils. An instant after that, there was a roar from his musket and a tremendous kick back into his arm. He heard the rattle of other men firing beside him. Smoke belched out in front of the whole line.

Howell was gripped by what he had just experienced. He lowered his weapon, and looked at the smoldering match and the little curls of smoke still coming from the pan and the barrel. He imaged that heavy lead ball flying away at tremendous speed. He was filled with awe at the power that he now held in his hands.

# Chapter X

## Reaching for the Heights

Porca miseria!" Giovanni flopped down on a rock beside Howell, filling and emptying his lungs in great gasps. They were nearing the top of a pass over the great Alps, but still he looked up at the jagged peaks towering around them. When he had gotten his breath back somewhat, he muttered, "I have dragged that sow of misery..." *whew* "...all the way up this damned mountain..." *uh* "...and I am about to tell her to walk." Then he turned to more ordinary things, and examined his shoes to see how they were faring.

The wind rose, and fell in maddening gusts. The citizens of Verona might still be enjoying the heat of the fading Summer, but everyone here was wearing his cloak, and had his back to the wind. The slopes above and below the friends were covered with the other soldiers of Mansfeld's Regiment looking for a place to take their hourly break. Even with their musketeer kit in the wagon to relieve them of that burden, all of them were panting deeply.

Beyond them, the great bulk of the Alps loomed, majestic and threatening. A wisp of snow blew off one of the summits.

A lad who spoke very little German sat down between Giovanni and Howell. He held his hand horizontal, then tipped it up at an angle, and then tipped it even more steeply, illustrating the ever-increasing slope of the trail. "Is more, more." He sounded amazed.

Howell nodded at him, wordless, with his chest heaving. There was no sign of the top of the pass. Worse, when they finally crossed over, they would have to wrangle all these wagons down the other side, which might be even more difficult than getting them up there in the first place. There might even be more than one ridge to cross, as well. He could not be sure about that, but between the shoulders of the mountains closest to them, he could see other peaks in the distance.

The whole of the Regiment was scattered in whatever places of rest they had happened to find, squatting and bundling whenever the wind blew. Their commanders dismounted to give their mounts a breather. These worthies gathered in small groups to discuss whatever it was that officers were always discussing. All up and down the road were wagons, chocked with stones behind their wheels to prevent them from rolling back down. The tarps across their tops bulged and fell and strained as the wind whipped by. The horses hitched to them hung their heads in exhaustion. Several of these wagons were ones that Howell knew very well were loaded with muskets and lead. One of them was the one that Howell and Giovanni had pushed every other hour all morning. Now it would their hour to just get themselves up the path, and let someone else do the pushing.

Giovanni glanced up. "Perhaps we are going to climb all the way to Heaven..." *gasp* "...and there will be angels to meet us at the top." He did not seem to believe his own suggestion.

"Heaven?" the lad between them shook his head. Then his entire little body shivered in the unexpected cold. "No Heaven. No, no."

"It could be worse," said another, older soldier. He had been having a particularly hard time of it. All the morning he had not been able to get enough air into his lungs, and he had only been able to keep up because Sergeant Blutworth had relieved him of duty. "It could be much worse." *Huh* "The last time I did it, …" *wheeze* "…it was snowing for days on end, and the wind howled like something from Hell."

*ah* "…that's what we're hearing." Giovanni gasped. "Now it makes sense. Rock Demons sent from Hell…" *gasp* "…to suck the wind right out of our lungs."

The youngster between them was breathing as heavily as any of them. "It like man has hand over mouth." He covered his face to show that he thought he was being suffocated by an invisible assassin.

Howell rubbed his temples. "The air is bad, but the part that is really hard to take is this headache I have got."

"Right now, I might be happier without a head," said Giovanni.

"It just pounds on and on. I wish there was something I could do."

"Water," said an old ranker. His voice was very faint.

"All right, all right." Howell patted his arm. "We will come to a stream presently, and you'll be able to drink." Howell thought the man was pretty selfish expecting them to supply him with water.

"Not me," the veteran wheezed. "Everyone. Lots of water. Stops the headache."

"Wait, wait, wait, wait." Howell turned to confront him directly. "You mean that I have this throbbing headache because I need to drink water?"

"Yes."

"If I need water, shouldn't I be thirsty?"

The old soldier shrugged his shoulders. "Have it your way."

Maybe this man was right. Howell had never done this before. Maybe he should pay attention to this experienced old hand. He leaned his head far back, stretching his muscles. He glanced down the slope. "God's wounds." It was Sir Horace riding up towards them. Memories of this man's betrayal overwhelmed him. Howell's hatred of Sir Whore's Ass blinded him, and filled every corner of his mind. His revenge was close at hand. He jumped to his feet.

Giovanni saw Howell stand, and then looked down the slope at what had caught Howell's attention. He had had only the briefest of glimpses of this Sir Horace fellow, but from the blonde hair and Howell's reaction, the man riding towards them had to be him. Giovanni stood up beside Howell. "This is a very bad thought, amico." No joking now.

Sir Whore's Ass grew closer and closer, his horse's hooves clattering through the loose stones. Could Howell kill him now? Was Giovanni right? Howell glanced up and down the slope. There were soldiers everywhere. If he did something obvious, like attacking the man with a knife, Howell would be cut down instantly. Perhaps he could manage to get the traitor's horse to fall, and do for Horace that way, but Howell just could not get his mind to focus on the task. He could not stick out his foot, and trip the horse as he had tripped the guard in Genoa's prison. What could he do? His thoughts were just not connecting in this thin air.

"Let us sit, amico."

Howell looked at his friend, and looked at Sir Horace. It took all the resolve that he could muster, but he sat. He pulled down the brim of his hat. He did it as much to keep from having to look at the object of his vendetta as to prevent Sir Horace from recognizing him.

Giovanni sat beside him and just hung his head so the brim of his hat covered his face. He looked like he was just sitting there, exhausted, which was not far from the truth.

Horace came towards them. His head was down, and the whole of his body was nodding with the rhythm of his mount's steps. He was almost upon them. He was exactly opposite them. Howell saw that hated profile, and thought of bloody revenge.

Giovanni placed his hand on Howell's thigh. He murmured to him, "Be calm amico. Now is not the time."

Sir Horace's head suddenly snapped up as he recognized his special wagons. He smiled as he reached the first of these. He rode on to have a look at the rest of his convoy. He never paid the slightest attention to Howell nor Giovanni.

When he took his eyes off Horace, Howell saw something completely unexpected. There was an elegant coach making its way slowly up hill. It was making better time than the wagons because it was relatively light, but now it stopped to give a rest to the horses that pulled it and to the men who were pushing. It was completely out of place in this workman-like convoy. Rosy had said that she and her company would cross the Alps. What better way than in such carriage in the protection of an entire Regiment of soldiers? Was it her? He could not see inside because the windows were covered with leather curtains to keep out the wind.

A shout came from somewhere downhill, and was taken up by all the Corporals. "Up," they called, their exhaustion evident in their voices. "Bedtime is over. Everybody on your feet." Their own dear Corporal Vlok was walking up the road, encouraging his men with his stick and his boots. "On your feet, bitches. We've got a mother big mountain to climb. Let's go. Let's go."

The lads reluctantly got to their feet, and prepared for another hour of thinning air, cutting gusts, and steepening road. Vlok reached Howell and Giovanni, and pointed at one of Horace's wagons. "Davidson, del Merico, you lot get back to work."

Vlok had said Howell's name rather loudly. Howell looked to see if Sir Horace had heard, but he saw no evidence that he had.

Vlok fetched a sharp blow across Howell's shoulders. "Now."

Howell snapped around, and was ready land a punch.

Vlok grinned back wickedly, daring him to disobey orders.

Sergeant Blutworth watched all this, and did not seem to be pleased by what he saw. There were plenty of other men who had never had a go at these giant loads, yet Vlok was making Howell and Giovanni take two turns in a row pushing one of these monsters. Blutworth wandered over.

Howell shifted his gaze from Vlok to Blutworth. "Begging the Sergeant's pardon. Could Giovanni and me ease off a bit? Maybe push that light carriage over there?" Howell was stepping well over the line by questioning his orders, and appealing to Vlok's superior.

The slightest smile appeared around Blutworth's lips. "That sounds fair."

A far more noticeable frown appeared on Vlok's face.

Howell started towards the carriage, but not towards the rear where he would do his pushing. Instead, he aimed at the coach's door. He still could not see through the leather curtain, but he resolved to take a chance. He knocked on the door.

The leather flap that covered the window came up, and a face emerged at just about Howell's eye level and only inches from his nose. It was one of Rosy's attendants. *Oh!* She recoiled in surprise. "Oh, Sir. You gave me quite…" and then she recognized who had caught her off guard. Her jaw dropped and her eyes grew wide. She sat thus for just instant, and then pulled her head back into the carriage. Howell heard women talking quickly, the carriage rocked as people inside changed places, and then Rosy's head appeared at the carriage's window. "Oh, Howell. Howell, you're safe."

"No thanks to our friend, Sir Whore's Ass."

"Horace?"

"Yes, he pretended to lead me to safety after that shooting, but then betrayed me to the guards."

Vlok slashed Howell with his stick. "You got the cushy job you wanted, and you cannot even be bothered to do that." He raised his stick menacingly. "Get back there," he said, pointing at the rear of the carriage.

Rosy looked out at the scenery, so that her conversation with Howell wouldn't be suspected.

Howell joined Giovanni at the back of the coach. Howell had laid out four paths that Sir Whore's Ass might take, and now two of them were eliminated. Horace was not taking his muskets to the Emperor. Whether he intended to use them to support the Revolution or to betray it was still up in the air. Horace's choice to go to Fredrick's court gave Howell hope that he might yet be able to get close enough to Whore's Ass to conclude his vendetta. Even if he did not kill the man, Howell was now close enough to work with Rosy in dealing with him.

Vlok raised his voice so that Howell and Giovanni at the rear, and the driver at the front could all hear him. "Let's keep the ride smooth, shall we? Ready!"

Howell swiftly gripped the carriage.

"Now!"

Howell pushed along with Giovanni, the horses took up the slack, and the coach got into motion. You could not exactly describe its movements as smooth, the road itself was a layer of loose stones after all, but at least the carriage moved steadily forwards without the jerking stops and starts of the wagons.

With Mansfeld's Regiment to provide an armed escort, Sir Horace had finally gotten his wagons on the move. It helped him cover what he was really doing with his cargo by including plenty of travelers who had nothing to do with his shipment of arms. He might suspect that Rosy had connected him to the assassination attempt as well, so keeping her close made sense.

Howell pushed for the better part of an hour, but finally break time came. Howell walked to the side of the road where he could get a good view of the carriage's window. Rosy looked out, trying not to be too obvious of what she was looking at.

Giovanni sat down next to Howell. He noticed that Howell was looking at the woman peeking out of the carriage window. "You keep staring like that, you are going to set her hair on fire. Oh, wait, her hair is already on fire."

"What? Oh." Howell turned to his friend. "That's Rosy."

"Who's Rosy?"

"Red hair. Pirates. Tobacco." These were all things that Howell had told Giovanni about while they were on the road to sign up to be soldiers.

"Rosy, eh? Could not you come up with something more interesting, like Apple-Head? Strawberry-Noggin?"

"I am glad I didn't." Howell shifted his attention, and looked carefully up the mountainside. Sir Horace was well forward, checking one of his wagons at the head of the column. There was nothing to say, though, that he might not come back at any moment.

"Time's up. Break's over." Vlok poked a few stragglers, and strode up to Howell, an unsettling smile on his face. His ears and nose were red with the cold. "I am putting you back where you belong." He looked at Howell's last assignment. "Nothing like that pretty coach," he sneered. Suddenly he was tough as leather again. "Get up there to that wagon." It was one of Sir Horace's dreadfully heavy wagons, of course.

Not only was the wagon they were given going to be a strain to move, but there was always the possibility that Horace would stop by for a spot of sight-seeing, and discover Howell. There was nothing for it, though. "Yes, Corporal."

"There's a good lad." The Corporal then pointed some other musketeers to the carriage. "You two get ready to push that." Vlok started up the slope assigning other victims to other wagons further up hill.

The established routine until now had been that men alternated between pushing, and simply climbing. Being told to push twice in a row, and now three times, was unfair, but that was undoubtedly exactly what Vlok intended, was not it? Howell scowled as he watched Vlok's back recede. "Bastard."

Giovanni positioned himself on one side of the wagon's tailgate. Howell took his place on the other side of the back of the wagon. There was barely enough room for both of them to push. Howell found himself at the very corner of the wagon box, both hands on the tail gate, and looking down the side past the wheels to where the horses stood. "Ready. It is heavy."

"No shitte."

The wind rose, and fell. It was impossible to anticipate when an icy blast would hit.

Wagon drivers up the slope yelled at their horses, and cracked their whips. There was a jangling of harness, and the sound of iron horse shoes and iron tires clashing on the rocks that formed the road. The sound got closer and closer as spaces opened up, and the next wagon in line had a place to go. It would be Howell's turn soon. He knew how bad this job was going to be. He had seen the muskets and lead that Sir Horace's wagons contained.

*Jee-ap!* Their teamster got their horses into motion.

Howell glimpsed the harness tightening as the horses took up the strain, and then he concentrated on adding his strength to the effort. It was like he was pushing against one of the mountains that surrounded them. The wagon wouldn't move an inch.

Giovanni put his shoulder against the wagon and re-doubled his efforts. Howell did the same. It was incredible that the wagon wouldn't budge. He could hear the horse's hooves scrabbling in the rocks, but they still weren't moving.

*Ahhh!* Giovanni's feet thrashed in the stones as the wagon backed into him.

Howell planted his feet, and jammed his shoulder into the tailgate again. The horses must have caught their footing at the exactly same moment, as the wagon suddenly reversed course, and bumped forwards a step. Then it moved again.

"Mio dio, it feels like this thing is filled with lead." Giovanni shifted his position so that he was not pushing with his shoulder, but with his hands.

As the wagon got underway, Howell found that he could sometimes push hard enough with his hands, but as often as not, had to resort to his shoulder. Sometimes he even had to put his back against the tailgate, and push with both his feet. It was exhausting. He and Giovanni coordinated their movements so they had make better progress. Sometimes it was "And…now!", and at others "Hold!" The wagon crept forwards, or lurched forwards, or hung up on a rock, or rolled easily on a flat patch of ground, and all Howell and Giovanni could do was try to keep up with its moods. Each time it landed heavily, Howell could hear the *thunk* of lead bars rising and falling on the bottoms of the crates.

The wind stopped for a moment, but then it roared again, and the tarp on the wagon in front of them just peeled back. They were about ten yards behind that wagon, but Howell could make out the boxes with the symbol burnt into them that he had seen when they were transferred from the ship to the wagons in Genoa. He was not surprised to have his suspicions confirmed, and it was easy to pretend that he had not noticed anything unusual.

The driver of the wagon that Howell was pushing set his brake. "Chock the wheels," he yelled. Howell and Giovanni found some rocks of the right size, and put them behind the wagon's wheels to prevent it from rolling downhill. The driver jumped down, and went forwards to help get the tarp back into place. The teamsters from ahead and behind hurried to help as well.

They were not the only men to take an interest in what was happening. Sir Horace came down the slope, his horse skidding through the stones, and Horace leaning so far back that the horse's tail nearly brushed his blond curls. "Break! Soldiers clear the road!"

"Break!" yelled the Corporals. "Stand down!"

There had been plenty of problems before, blockages, that sort of thing, but the men had always just stood, and waited for it to be sorted. Not this time. Their rest period had come much earlier than was normal, as though Sir Horace did not want anyone seeing what was in his wagons. Howell knew why that was, but could he seriously think that one of these common soldiers might be able to cause mischief with the knowledge?

Howell led Giovanni away from the trail keeping his back to all the bustle around Sir Horace's wagon. Howell was calm enough, and sufficiently exhausted, that he admitted to himself that this was not the time to try and visit justice upon Sir Whore's Ass, hard as that was to admit.

The mountains still looked down upon them from either side, but the convoy was almost certainly approaching a saddle of some kind. Perhaps in the next hour or two they would reach the top of the pass. He saw, too, that Rosy's carriage was just uphill from him. Her coach had been passing the wagons the last time he had been able to look, but now the road was too narrow. He saw Rosy's head protruding from the carriage window, looking sternly at Sir Horace.

Giovanni got his attention and pointed towards a trickle of water flowing down beside the road. "Let's put that headache remedy to the test." Giovanni gasped in the thin air. "In any event, I could use some water."

The mates found a place along the bank among the crowd of soldiers with the same idea that they had. Bank was not really the right term. The brook babbling downwards had margins that were formed only by wetter and drier rocks. Howell found as good a location as was still available, and shoveled water into his mouth with his hands. The water was terribly cold, just this side of freezing, so his hand quickly went numb, and his teeth hurt from drinking it. He put his hand in his armpit to warm it. The road ahead seemed to be pitched even more sharply than what they had barely coped with until now. The next stretch was going to be tough. He looked significantly at Giovanni. He leaned over to whisper, "I saw the crates in that heavy…" *whew* "…that heavy wagon when the cover blew off just now."

"Amazing! You mean to say…" *gasp* "…we're not moving a train of empty wagons?"

Howell was not in the mood for one of Giovanni's cocky responses. "I had to count every one of those bloody…" *uh* "…great boxes as they were loaded on and off the ship."

"So what does that mean?"

"It means that the boxes are filled with muskets... *whew* ...and bandoliers and lead..."

*Ahhh* Giovanni registered understanding of why the wagons were so heavy.

"The fellow who had me thrown into the cells with you, and tried to kill me, you remember, at Genoa?"

"That scassacazzo tried to kill me, as well, when we ran out through the gate."

"...well he's the one taking these weapons across the Alps."

Giovanni glanced at the heavily laden vehicles. *Uh hunh*

"And I think I know what he's going to do with them once they are on the other side." Howell looked down into the valley. It irritated him that he was helping to transport Sir Horace's goods, but so long as he was close to the muskets and lead, he would remain close to Rosy as well. Once they were across these mountains, she would talk to Elizabeth, and everyone would discover exactly what Sir Horace had planned for these arms. There was also the possibility that he had cross paths with that murderous Sir Horace at some place a bit less public than a road going over a pass in the Alps.

"Up. Up." The Corporals began to call out. "Fall in."

Howell stood and there was Sir Horace right before him, looking uphill at his string of wagons. Howell put his hand on his knife's hilt. Now was the critical moment. If Sir Horace recognized him, Howell would strike. He would get his revenge, but after he had attacked, he would have only a tiny chance of survival. More likely, though, Sir Whore's Ass would remain out of reach, and set the soldiers upon him. Then Howell would have to choose between cutting up his mates, or else escape by running up the rocks and gravel covering the slopes on one side or the other of him. He was not sure which way he would pick. The only certainty in this was that he wouldn't take the coward's way out. He wouldn't let himself be taken without a fight.

Then Sir Horace was gone. He rode up the slope to a place beside the wagon with its newly attached tarp. He seemed to take no notice of the wagon he had just left, nor of the soldiers leaving the stream to resume pushing his precious wagons.

Howell pressed his knife back into its sheath. He followed Giovanni to the rear of their wagon.

"Move out, you scum." Vlok announced the continuation of the hour's duty.

The friends resumed their positions and, when the driver whipped up his horses, they resumed their pushing as well. Howell had seen that the road was becoming even steeper than it had been, and their labors were soon only barely adequate. Then they came to an obstacle that they simply could not overcome.

Corporal Vlok came striding down the hill. "What's going on here?"

Howell thought that this was so obvious that even someone as dim as Vlok could make it out. It is a bloody wagon that's too bloody heavy to bloody move, that's what it is.

"Push, you lazy pricks! Push!"

Both Howell and Giovanni had their backs against the tail gate, and both feet planted, but they could not budge the wagon.

Vlok seemed, at last, to understand that something more was needed. He grabbed the sleeve of a young musketeer who had the bad fortune to be walking by, and flung the boy towards the back of the wagon. "Push, damn you! Push!" At the same instant, he fetched Howell a sharp blow across his chest and face.

Howell started in surprise. The musketeer blundered straight into him. Suddenly the forces on the wagon were different. Without the pressure that Howell had been providing, the wagon lurched backwards, and crashed into him. He lost his footing from the impact, and suddenly the wagon was rolling down the mountain. Howell dug in his heels, but without effect. His shoes plowed two furrows in the stony road, but the wagon picked up speed. The driver finally realized that what was going on was not the usual ebb and flow of his wagon's movement over the mountain. He set his brake, but it hardly made a difference.

Struggling to regain his balance, the musketeer that Vlok had thrown at the wagon's tailgate now found himself directly in line with one of the rear wheels. The wagon box struck him and knocked him straight back. An instant later, the wheel rolled right over him, crushing legs, torso, and finally his skull. His shriek of terror was still echoing from the slopes of the mountains when every man who could get a hand on the wagon finally brought it to a halt.

Bit by bit, Howell slowly relaxed, but only physically. The fury inside his heart remained as hot as a fever. He brought one foot under him, then the other, and gently, tentatively, took his back off the tailgate, and stood. He looked around for that son of a bitch, Vlok. Was he just an idiot, or had he truly hoped that it might be Howell who would be crushed into the road by the wagon?

"Clear the way there. Let me see!" It was Rosy hurrying down the slope with her attendants right behind her. "Clear the road." She was holding up her skirts so she wouldn't tread upon the hem. Concern marked her face. She almost lost her footing on the rocks as she came down, but finally reached a place where she could get a good look at Howell. She let the air out of her lungs in a great sigh of relief.

Howell was not relieved. He was shaking. He was shaking from his exertions, but he was shaking with rage, as well. There was a dead man right beside Rosy. This was Vlok's doing. He had caused a man's death with his pride and incompetence, and now he was nowhere to be seen. Howell looked down at the poor fellow crushed into the stones.

Rosy followed Howell's gaze. "Oh, my God." She put her hands to her mouth. *Oh* She turned away, fighting the urge to vomit. *Oh* Rosy's women put their arms around her.

Howell heard something, and turned to look uphill. Vlok was bustling down towards him. Howell could see his mouth working as vigorously as his stick, and presently some of the words became intelligible. "God's blood! What's going on here?" He was thrashing wildly with his stick at anyone he encountered. "Move! Get this wagon moving, you lazy swine! Move! Move! Move!" He picked several men at random, and shoved them towards the tailgate of Howell's wagon. "Push, you maggots! You lazy slugs! Push!"

"Stop!"

Everyone on the mountainside froze at the sound of a woman's voice.

Rosy gathered herself, and pointed at the corpse ground into the road. "You will stop what you are doing until this man has been properly buried."

"Buried? Buried?" It was as though Vlok had never heard the word before. "My job is to get this lot up this pass, and by God, that's what I am going to do! No one is going to say that Corporal Vlok did not do his duty."

"And right now, your duty is to see that this man is given a proper burial." Rosy confidently walked behind the wagon to ensure that no one could approach the tailgate without feeling her wrath.

Dozens of soldiers crowded around the confrontation, eager to see the show. Vlok snarled and flexed his hands. Howell stared at him. This creature had caused a man's death through his stupidity, or perhaps villainy, and was now acting as though it was only a matter of traffic flow.

Rosy cast her eyes around the crowd challenging anyone to disagree with her. Her particular attention was focused on Vlok.

He grit his teeth, and strode up to Rosy. "Clear off, Missy, so men can get to work."

"The only work I want to see performed here, you heartless scoundrel, is for you to see to it that this man is buried." Rosy's eyes never wavered.

Vlok's eyes were fixed upon Rosy's in turn. He hesitated to use his stick upon a woman, but he could not allow his authority to be challenged by a woman, either. He turned to the men behind him. "You. You. Get up here, and start moving this wagon."

One of the men hesitated. The other walked up towards Rosy, but stopped when she blocked his way.

"Your comrade wants burying," Rosy said quietly.

The man stopped, and considered. The man under the wagon was his comrade. This man had shared his initiation as a musketeer. They had worked together to work out how to fire a musket despite Vlok's attempt to confuse them. Did he owe more to his wretched Corporal, or to his friend?

Vlok charged at him, viciously slashing with his whip. The poor fellow held up his arms in an attempt to ward off the blows. "That's a flogging for you! No piece of filth…"

"Stop this nonsense." Rosy stepped in to separate the two men. Emphatically, she said to Vlok, "Bury this man. Now."

"What's going on here?" Sergeant Blutworth came limping up to the scene of the blockage. "Corporal! Report!"

Now Vlok's eyes focused on Blutworth, but he continued to stand where he was. He was positively twitching, and took a nervous glance at Rosy.

"Now!" Blutworth stopped a few paces downhill of Rosy and Vlok. "Report!"

Vlok began to walk towards his Sergeant, but before he could take a second step, Blutworth shouted, "Halt! Report!"

Vlok jerked to a stop. He touched his hat brim in respect. "Sergeant, this man…"

"What?" Blutworth shouted loudly enough that he could be heard all over the mountainside. Quite a crowd had gathered to observe the spectacle. "Speak up man."

"Sergeant…"

"Louder!"

"Sergeant, this man disobeyed my orders!" Vlok clenched his teeth, and pointed at the latest victim of his stick. The hatred fairly shown from his eyes.

The crowd was deathly silent. Every man wanted to hear this exchange.

"Well, son? What order did you disobey?"

Rosy stepped forwards. "He disobeyed the order to leave this man lying here to be trampled into the road rather than give him a proper burial."

A murmur swept through the crowd.

Blutworth craned his neck to look around Rosy, and see the poor fellow's remains. His face grew very stern. "Is this correct, Corporal?"

Vlok brought himself stiffly to attention. He did not look exactly at Sergeant Blutworth, but rather somewhere over his head. "My orders were to keep this column moving and…"

"Moving at the cost of a man's life?" Blutworth's voice rang out angrily.

Vlok remained stiff and silent. He pursed his lips and took a deep breath.

By this time the hushed spectators included everyone who could crowd into the little amphitheater. Even some men who had already passed this point had doubled back to take in the performance.

"Let me be certain that I understand the situation." Blutworth kept his eyes fixed upon Vlok, but roared out his statement. "This lady here…" Blutworth touched the brim of his hat and nodded towards Rosy, "…brought to your attention that one of the men in your care, one of our comrades, had been killed, and that you needed to bury him."

"Yes, but…."

"No 'buts', Corporal. Did you or did you not bury our comrade?"

"No, Sergeant!" Vlok boomed his answer. The Sergeant was making him deliver his replies in a voice loud enough to be heard by everyone in the audience up and down the mountain road. "I will see to it immediately, Sergeant."

"You'll see to it when I dismiss you…" Blutworth's voice boomed over the crowd. "…and I will dismiss you for good and all. You are no longer a Corporal, Mr. Vlok, you are now a private soldier." Blutworth looked around at the crowd. "Davidson. Where are you?"

"Here Sergeant." Howell could not believe what he was sure was about to happen. He stepped through the crowd, and saluted his Sergeant.

"You will take Vlok's place as Corporal. As your first duty, I want you to see to it that Vlok deals properly with our comrade's remains."

Howell beamed with pride and malice. "Yes, Sir."

Giovanni grinned with amusement.

Rosy smiled at justice done.

Vlok scowled in the road, exposed for all to see, mute, and alone. Even men who had supported him before now abandoned him.

The crowd began to sense that the show was over, and that it was time to get back to business. If the Sergeant had waited any longer, he would have lost them, but his timing was perfect. "Sooner than you might want to think, men, we will be going into battle…" Blutworth looked around at the sea of faces turned towards him. They were new men mostly, recruited in small towns and from farms along the route that the Regiment had marched. They had never ventured so much as ten miles from their homes until now. They had certainly never smelt gunpowder in anger. They had never been asked to stand fast as bullets swept through their ranks, or great masses of cavalry had charged down upon them. "…and when that happens, each man present today will be holding his life in his hands."

The truth of the Sergeant's words gave pause to his audience. Some of the boys looked nervously at one another. Some swallowed hard, and stared off into space.

"We will be alone, and facing men who want to kill us. When that day comes, where will you find your courage? What will give you the strength to face danger with confidence and bravery?" Blutworth looked pointedly around at the men gathered to hear him. Their eyes were riveted to him. "There is one way, and one way only, that we will taste victory. The life of each and every one of us will depend upon the determination of the men on either side of him. The determination of his comrades to do their duty."

He had the crowd gripped in his fist now. Their attention was focused again, and they were shifting and muttering and watching him with bright, hungry eyes.

"This woman stood with our comrade!" Blutworth pointed directly at Rosy. He paused to let his words and his gesture seize them. "Can you do less? Will you do as well? Will we stand by one another and live, or we will desert our comrades and die? Will we endure hardship and fear and wounds for one another, or we will think only of ourselves? What will it be? What say you soldiers of Mansfeld's Regiment? Will you stand with me?"

The men were so new, so unsure of themselves, and so hypnotized by what was happening around them, that no one took up the Sergeant's challenge.

Howell sensed that something had to happen, and happen this very instant. He snatched his hat from his head, and waved it in the air. "I will stand with you Sergeant!"

"And I!" The crowd burst into turbulent shouting. "I will stand!" "You can count on me!" "Yes!" "Together!"

Blutworth's arm swung majestically to point up the road behind Howell. "Will you swear by this dead comrade that you will never desert your flag?"

"Yes! Yes!" The crowd seethed and yelled. Hands and hats were held aloft. "I swear! I swear!"

*Oow! Oow! Oow!* Howell's bark swept over the clamoring assembly and seemed to echo faintly from the great peaks that loomed over them. He was overflowing with pride. He turned to look in admiration at Rosy, his beautiful and brave Rosy.

His joy had one flaw, though. Now his name was in the open. Blutworth had appointed him. He had replaced Vlok. His Captain would certainly want to know who had taken the place of his toady. The Captain had no idea of the connection between Howell and Sir Horace, though. How long would this last?

# Chapter XI

## Sir Horace's Advice

Sir Horace sat in an antechamber in the castle in Heidelberg. This ancient structure held the throne of the Count of the Palatinate, Fredrick V. This man was currently the spokesman for the disorganized Protestants in Germany. The only thing that held them together was their opposition to the Emperor's desire to return all of his subjects to the Catholic faith, but they had no idea of how they should go about it.

Horace was going to give them a course of action. He would show them a path that would ultimately lead them straight into the trap that he would set. This trap would put them into the Emperor's hands, and Horace was going to use Count Fredrick to lead the way. The first step, then, was to get Fredrick into his grasp.

There were complications, unfortunately. Davidson, that nosey bastard who was always dragging around that freckled-faced strumpet, had gotten a look inside Horace's boxes of musketeer kit. He had seen the muskets and whatnot, but had he noticed that there was no powder, and had he realized how important that fact was? Horace had tried to silence him by capturing him, and turning him over to the Genoese for hanging, but Davidson had vanished like smoke in the wind. Where was he now? Had he told that red-headed harlot what he had seen?

Worse, had Davidson's trollop told the Countess what the boxes contained? He was going to pretend that he was an agent for King James of England, yet the Countess would be very suspicious about this. She knew that James did not want to get involved in a war in Germany. He had to be sure she knew as little as possible so that she could make as little trouble as possible.

The door to the reception hall swung open. A formal servant bowed deeply, and swung his arm to invite Sir Horace through the magnificent doorway. "The Count will see you now, Sir Horace."

He found the Count standing in a large room decorated with every sort of luxury one could imagine. There were huge tapestries upon the walls. Silk curtains covered the windows. There were delicate porcelain statues and jars, all of which had been imported from China at outrageous expense. Lavish furniture sat upon ornate oriental rugs. The Count greeted his visitor, "Ah, Sir Horace. How nice of you to visit us."

Horace was not here to admire the scenery, however. He had a job, and he plunged into it. "It is a great honor to be received by you, your Lordship."

"Nonsense. It would be churlish of me to refuse an audience with a man who has crossed the seas and the Alps to speak with me."

"Thank you, Count. It has indeed been an exhausting journey. Men died defending our ship in a fight with pirates. Others died amongst the mountains. I myself barely escaped death in an avalanche." It never hurt to add a little lie to boost one's image, did it?

"Oh, my goodness." The Count was shocked. "What is it that has caused you to face such perils?"

Horace composed himself. This was the moment. "Count, you have certainly heard of the attempt on the life of the Duke in Genoa."

"Of course. A shocking incident to be sure."

"I do not believe that this was the act of some random madman. In my mind, there is no doubt that the man who fired the pistol was an agent of the Emperor."

"What?" Apparently, this notion had never occurred to this dim Count Fredrick.

Horace was pleased to find that this man was every bit the gullible fool that he had been led to expect. "This is a direct challenge by the Emperor to all the Protestants in Germany. As the leader of these Rulers, I humbly suggest that you must respond to this threat."

Count Fredrick mulled this over. He looked several times out of the corner of his eye at Sir Horace who did his best not to stare at the man. Finally, Fredrick spoke. "What you would suggest?"

"Sir, the Emperor has no army of his own. Even as we speak, he is in negotiations with the Catholic states in Germany asking that they join together, and come to his aid. If your supporters had a strong army, the Catholic states would be reluctant to go to war."

"That is all well and good, Sir Horace, but we have no such army."

"I can help with that, Count. You have asked what caused me to make the difficult trip over the seas and over the Alps. The answer is that I have brought to you muskets, bandoliers, and enough lead to make thousands of balls. With the arms that I have brought, you will need nothing more to create the most powerful Army in Germany."

Fredrick's jaw went slack and his eyes showed the astonishment that he felt.

"It is no secret that King James of England has openly opposed your Revolution, but the attempt on the life of your ally has changed the King's mind. James sees that the only way to stop the Emperor is to support your efforts. Therefore, I come before you as the Special Envoy of King James to present these munitions to you so that they can be used to further the cause of the Revolution."

Fredrick beamed at the opportunity offered to him. Sir Horace's gift would make him not only the unchallenged leader of the Protestants, but the supplier of the weapons they needed to make the Protestants a force to be reckoned with. Horace was offering him both political and military power of which he had only dreamt until now.

"Naturally, he wants his change of heart to remain a secret, so I must ask you to keep this fact well hidden. I will even ask that you to keep the Countess Elizabeth in the dark." Horace became more serious. "She has taken odd positions on some of these matters, and will only confuse things." Then he lightened again "As women are wont to do, eh?"

"Oh, yes," laughed Fredrick. "Women will talk, won't they? But we have men's business to attend to, Sir Horace, and you have risked all to provide the tools that are so desperately needed. You are very kind, Sir Horace. Very gallant." Count Fredrick searched for words. "Such praise does not come close to expressing the depth of my gratitude, Sir Horace. It barely hints at the debt owed to you."

Horace had every intention of making good use of that debt. "Then I bid you good day, your Grace."

"Certainly, certainly." Count Fredrick was lost in dreams of himself as the leader of a great host coming to the rescue of Princes all over the land. "Certainly, Sir Horace. Certainly."

Horace stepped out into the antechamber. The butler handed his hat to him. As he took it, Horace glanced to one side. He was astonished to see that freckle-faced hussy walking up to him.

"Ah, Sir Horace. I am so sorry to have missed you."

"Nonsense, Miss Kincade. The Count and I were just exchanging pleasantries." Had she been eavesdropping? Had she heard everything? He could kill her, and then it wouldn't matter what she had heard. However, the Countess already suspected that King James had not changed his mind at all, and that Horace's story was a pack of lies. He did not have much time to get control over Count Fredrick, and he could not afford to waste what little time he had on a fight with the Countess. If he killed this red-headed jezebel, the Countess would know that all her suspicions are correct, and she had work tirelessly to thwart him.

He looked at this Kincade woman. He had to let the wench live.

# Chapter XII

## Payday

It was a fine Autumn day as the long, rambling column of Mansfeld's Regiment moved steadily towards south Germany's great city of Nurnberg, its spires now plainly visible ahead. The road they marched was broad and level and seemed like a blessing from Heaven after their exhausting journey up and down the steep paths through the Alps. Even the dust raised by their feet was a welcome change from the rocks and wind (and once, even snow) of the great mountain range.

The column was led and trailed by half of the Regiment's fifteen hundred musketeers and accompanied by the constant, gentle clattering of their chargers. The mass of the Regiment's spearmen formed the center of the column. It was far too fatiguing to carry their enormous pikes erect for any significant distance, so they 'trailed' them, holding onto the business end with its iron point, and letting the butt drag in the road. This made for an extremely long column.

As it happened, the shotte of Howell's Company were near the head of the Regiment, so that Howell and his mates saw the city clearly, while those further back could only see it half-obscured by the dust they raised as their march proceeded. Howell reached the ditches and bastions of the earthworks that formed the line of defense for the city. The city's great gate stood open before him. The musketeers passed through and into a goodly-sized street lined with the sort of half-timbered buildings that Howell was used to seeing in London, rather than the white washed stone he had seen in Italy. There were curious onlookers on either side of them, and some of the citizens stood in their doorways to watch or leaned out of their upper-story windows. There were many smiles and quite a few waving hands. Some of the citizens waved other objects as well. Many of these 'other objects' were bread or mugs, so that Howell quickly realized that these people were not welcoming an army to defend them, so much as trying to attract some of the business that would undoubtedly be conducted that night.

For some reason, the old-fashioned notion lingered that the spearmen were the senior members of an Army, and that therefore they should be paid half again as much as the musketeers, and placed in the center of any formation. Presently, these trailed their pikes through the gate. Of course, it wouldn't do for these esteemed warriors to display their weapons in such a workmanlike formation as they passed the reviewing stand. So the whole of the Regiment would have to stop to allow the pikemen to get their weapons vertical and presentable to their employer. Meanwhile, there was nothing for the musketeers to do but cool their heels, and await the order to march.

Giovanni shifted his heavy musket from his shoulder to put the butt on the ground. "Payday." *Hunh?* Giovanni poked his elbow in Howell's ribs. *Hunh?* "How many sacks of money will each of us get, do you think?"

"I am sure I do not know, but I doubt it will be sacks," said Howell. "Still, I have never seen it done."

"I have." One of the old sweats smiled at the friends. "It is not difficult. Deciding what to spend your money on afterward, that's the tricky bit."

"I do not think it will be tricky to find wine," said Giovanni.

Howell had another objective. "Some decent food will be welcome."

Their old comrade was not interested in either of these observations. He was lost in his own world, imagining what was soon to come. "There will be some of everything to buy!" He smacked his lips. "Beer and sausages and beer and women and beer and…"

Giovanni laughed and slapped his hand down on the veteran's shoulder. "And beer, am I right?"

The pikemen held their eighteen-foot spears pointing straight up, so it appeared that a grove of saplings was being shouted, and shoved into the desired formation. At last it was done, and Mansfeld's Regiment was ready to impress the grand personages upon the reviewing stand. A small man on a horse, the legendary Ernst von Mansfeld himself, followed by several aides, rode by, and placed himself at the head of the column. Their triumphal progress began. A brass band *oom-pah-pahed* happily somewhere ahead.

The broad thoroughfare upon which they were marching abruptly opened out onto a very large square. Centered on one side of this space was an elevated platform which looked like its more usual employment was as a scaffold for executions. Flying above it was a flag of two horizontal stripes, light blue and white. A gaggle of dignitaries stood upon this reviewing stand.

All the citizens of Nurnberg were cheering now, including the spectators in the windows surrounding the square. The band honked loudly.

"Eyes right!"

Everyone on the reviewing stand was talking and waving and generally making it a festive occasion. Giovanni hollered above the noise. "Mama mia, who's that standing at the front?" He was referring to an elegant young woman with her light-brown hair pulled back across her head with every strand in perfect alignment. She was wearing a blue dress that shimmered in the Sun, and this was topped with a double row of white beads around her neck.

Howell looked knowingly at his friend. "I'd lay money that you are currently gawking at the Countess Elizabeth of the Palatinate, in the flesh." The men around them stared at this rare creature, making impolite observations, and noises that might have embarrassed a hog.

As Howell walked by, his perspective changed, and he could see the dignitaries from a new angle. The Countess's dress seemed to sparkle. Then another woman became visible beside the Countess. This one was much more plainly dressed, but for Howell, the red hair was unmistakable. "That's Rosy!"

Giovanni's eyes left the Countess and registered who it was that Howell was fixed upon. "Rosy, eh?" Giovanni shook his head. "Every woman within walking distance of Merico had black hair. Or grey hair that had once been black. Since I have come on this journey, I have met people with brown hair and yellow hair and now red."

As befit her station, Rosy's dress was far less elaborate than her mistress's, but it was comely all the same. The toes of her shoes peeked out from beneath her long, black skirt. Her tight grey bodice was squared off to reveal the tops of her breasts, and her puffy sleeves seemed to be almost separate from the rest of her garb. Every tiny movement she made just penetrated directly into Howell's desire.

These fetching women were the best part of the show to the men. The shouting filled the square, competing with the music from the horns and drums, and with the cheers from the houses.

*Oow! Oow! Oow!* It was a spontaneous choice. Howell felt the same exhilaration and defiance that he had felt when he had overcome the pirates and survived Vlok's initiation, and the barking just came. Her face lit. Howell kept his eyes firmly upon Rosy as he marched past. She frantically scanned the crowd, looking for the source of his distinctive call, but she never made eye contact.

Then she was gone. Anything he might have heard was drowned out in the general commotion. Anything he might have seen was blocked as the column marched behind a barrier of horses and flags and officers with be-plumed hats. Below one of the hats was a mane of bouncing, blonde curls.

"God's body!" Howell twisted his head to ensure that Sir Horace did not see him. So close! So very close! It was all Howell could do to hold on to himself. He wanted to break ranks, dash over to the treacherous pig, and gut him then and there. He had like to exact justice right in front of this assembled multitude. He kept himself under control. Control.

"Porca misera! Isn't that...?"

"Yes." Howell grit his teeth, and clenched his fist.

They marched on for a moment. The noise from the rowdy crowd around the reviewing stand faded away.

Giovanni turned to his friend with a great smile upon his face. "You are a lucky man, amico mio."

"Lucky?!? How can you say that? Do not you understand what just happened?"

"Si, si." Giovanni nodded confidently. "I understand the ways of the vendetta, and I know that your problems are at an end."

"At an end? Have you lost your wits? What if that was my last chance?"

"That is not possible."

"What?" Howell was beginning to free himself from the blind rage that had overtaken him when he first saw Sir Horace. He was starting to actually listen to his friend, but Giovanni was speaking nonsense. Was this just another of Giovanni's endless jokes? "Why do you say that he cannot escape me?"

For once Giovanni was straightforward. "Because it is obvious that you are now under the protection of Saint Justin."

"Saint who?"

"Justin." Giovanni admonished Howell with his index finger. "The Patron Saint of the Vendetta."

"You have a saint for killing people?"

"Well, he's more like a Saint for giving people what they deserve, but it is the same thing, no? Still, someone keeps putting this man in your path."

Howell's brain slowed down enough to consider this. In truth, there was something uncanny about how he kept finding himself in the presence of the hated Sir Horace.

"You escaped him at the city gate in Genoa. You saw him again and again on the road over the mountains. Now you find him in the company of your Regiment's officers. You see? St. Justin is looking out for you." Giovanni put his hand on his friend's shoulder. "Fear not. Saint Justin will see to it that you have your revenge."

Howell was so not sure that he believed his friend's strange reassurance, but the moment had passed, and Giovanni had prevented him from doing something stupid. Perhaps that was all that he had intended. Howell looked dubiously at Giovanni. St. Justin? Really?

Their brief part of the ceremony at an end, Howell's Company turned into a street, and left the square. Mansfeld's Regiment began to move through Nurnberg's increasingly twisting streets. Welcomed on every side by the people anxious to sell, they marched out of another of Nurnberg's many gates, and into the open. The sky was again visible as a vast canopy instead of just a narrow blue strip between rooftops.

In the fields beyond the city was another city. This one was like the place in which Howell and Rosy had sold her tobacco. The word had gone out that Mansfeld's Regiment of mercenaries would be having a mustering for pay, so every vendor within driving distance had come to get a share of the silver. Tents and wagons offered to sell you anything that you might want. Meat was roasting, beer was flowing, and the smell of commercial activity was everywhere. It was not just the proprietors of the rolling brothels who knew the value of a pretty face, or an artfully displayed bosom, either. "Good beer here!" the comely bar maids called from behind the counters. "Cheese!" "Fortunes told!" "Shoes!" Men, women, old, young, vendors of every kind offered every article and service of which a man could dream.

Giovanni marveled at this glimpse of paradise. Even Howell forgot his brooding long enough to grin at his friend.

"Halt! Dress ranks." They had passed the wonderland, and now were in a field just beyond. "Straight lines, you lot!" Their Company's flag was planted in the ground before them and flapped lazily in the breeze. The Captain sat his horse between his flag and his men, watching them assemble. There was a table set up next to the flag, and Howell recognized the man sitting behind it. With his spectacles and his jaw thrust forwards and his muster book and writing implements arranged just so, it had to be the same Mr. Bell who had put their names into the muster book at Verona. Behind

Mr. Bell were two stout wooden boxes reinforced with iron, and next to them were two heavily armed, strapping lads to guard them. One of the boxes was firmly closed with a padlock the size of a man's hand, but the other was open. The aura of money radiated from them both.

The Captain addressed his men. "You see our flag before you. For those of you who are doing this for the first time, when I give the order, you are to stack arms and proceed in an orderly, and I emphasize that word, in an *orderly* way to the base of the flag to receive your pay."

A great shout went up at that.

"Fall out!"

Well, it might be called orderly compared to someone tipping over an ant heap, but there was certainly no time lost. The old sweats formed small groups with one man, designated by custom or bad luck, to stay behind as the guard for their weapons, while the rest moved off to line up in front of the flag. The queue formed in a few beats of one's heart.

Howell and Giovanni were left behind by this swirl of activity, and found themselves standing in the field with the other mystified recruits. Corporal Davidson did not hesitate to lead the way, however. He noted what was happening in other groups, and handed his kit to the boy next to him. "Congratulations sonny! The youngest always gets the shitte detail." Howell started off. Giovanni grinned sheepishly, but did as Howell had done before hurrying off. The lad who had been left in charge of the muskets pulled a face.

One of the men who had quickly reached the head of the line was intercepted before he could plunge into the carnival spread out around them all. A prosperous-looking gentleman with a face scarred by small pox took this man aside, and it did not look like a casual chat.

"It appears that this fellow has met an old friend." Giovanni spoke with evident sarcasm.

"About as welcome a 'friend' as the one that visits women every month." Howell caught a despairing look from the debtor as they walked by him. "His poxy chum is probably just as regular, too, showing up without fail for every payday."

"I know such men." Giovanni's usual wisecracking was completely absent. "He will greet you like a brother and give you silver. Today. But every month after that it will be harsh words and his hand under your nose demanding eight for every five that he was so happy to lend."

Just as Howell was about to take his place in line, Vlok and two ass-kissers he had found rushed in, and pushed their way in front of Howell and Giovanni. "This is how it is done, slugs." He leered at Howell, and then looked up at their Captain, mounted just before them. He took in the scene with a little smirk.

Howell let a few men pass him so he wouldn't have to stand next to Vlok. The line moved slowly. The wait seemed endless. One by one, each musketeer returned from the front of the queue. Some were carrying their hat upside down, with coins jingling in it. Some went by them with looks of contentment or resignation or even anger. Howell and Giovanni could not decide what all this meant. Two men with big smiles on their faces passed them laughing about how their first stop was going to be a beer garden they had seen in town.

Giovanni licked his lips. "I like the sound of that, but I wonder if I might not find some good Italian wine, too."

"I guaranty it, friend, but I will tell you what will be even better."

"Better than drinking?"

"You and I are going to find a nice tavern somewhere, one with all the comforts of home." Howell gave his friend an absurdly broad wink. "And we are going to have the best night of our lives. All on me."

"On you? Why?"

"Because of the way you saved my arse at the gate at Genoa."

*Awww…*

"It is true. I admit it." It was painful beyond measure for Howell to say these words. He had never met anyone who came anywhere near his level of resourcefulness, and acknowledging that his plan had failed, and that he had needed rescuing tore at his soul. "Those cutthroats would have found me out for certain if you had not started that ruckus with Sir Whore's Ass. That was the act of a clever man and a true friend, and it deserves to be rewarded."

Giovanni was positively blushing at his friend's tribute. "I was glad to help."

"And I will be happy to buy you meat and bread."

"And wine."

"Oh yes." Howell laughed. "Wine certainly."

"And a woman."

"Positively. One of those brawny mares that you favor."

"Next!"

Giovanni waved Howell forwards. "Collect your money now so that I may spend it tonight."

Howell stepped up to the table.

Mr. Bell adjusted his spectacles, stretched, and flexed his hands to loosen them. He picked up his quill, and rapped the table with his knuckles. "Right, soldier, it is your turn," he said in his Scottish accented German.

Howell stepped right up and did not wait to be asked for his name. "Howell Davidson, Sir."

Bell looked up. "The writer, as I recall." He had switched to his Scottish accented English now. He briefly shuffled though the stack of papers that stood before him, and snatched out one of them. He began reading. "Your wages are owed to you, minus the cost of your food and this month's charge for your musketeer's kit. You bought a new suite of clothes on the Company's account," and he looked dubiously at Howell's muddy, road-weary attire. Then he stopped and looked more closely at the chit. He looked up at Howell. "And you are being fined for pushing our Captain into the mud at Verona."

Howell had not pushed the man, but there was nothing to be gained by an argument now. "So how much do I get?"

"Nothing."

"What? Nothing?"

"Aye, Mr. Davidson. And you are 139 kreuzers and three pfennigs to the bad this month, and will be charged 116 kreuzers and two pfennigs in each of the next two months to defray the cost of your equipage. We shall just have to carry it over 'til next month."

Howell stared.

"You said you could do sums, man. Do you question my figures?" Bell placed Howell's chit in the pile. "Next!"

Howell's version of a payday was over. He turned and walked slowly back past Giovanni.

"Nothing?" Giovanni was stunned.

"Not a damned heller. And I still owe on the bloody musket and the damned bandolier."

"Next!"

Giovanni could not afford to linger. He patted Howell's shoulder as he proceeded to the base of the flag.

It seemed that Howell was the only man on the field who was putting on his hat as he left the paymaster. He had thought to boldly escape Genoa, and instead had to be rescued by Giovanni. He had thought to repay his friend's kindness and bravery, and yes, his superior cleverness, as a way to salve his own embarrassment, and instead he had been buried under another shitte-load of abuse. Penniless. Dependent. Howell was fit to gouge out eyes.

Giovanni joined him, and the two of them started off. He was wearing his hat as well.

"You got nothing?"

"Oh, I got everything I was owed." Giovanni was his irrepressible self again. "I got a bag of air. I got laughed at by the men behind me in the line. How could a man desire more?"

Howell grunted, and looked around at the camp. They walked by a booth offering magic charms, some which claimed to make one bullet-proof, others to protect against disease, or to attract lovers.

Giovanni turned to Howell. "Maybe we could borrow one of those charms long enough to turn these smells into something we can eat."

*Um hum*

Consumed with their own thoughts, Howell and Giovanni barged heedlessly into a group of soldiers determined to swiftly convert their pay into pleasure. "Watch it mate!" One of them swung a sausage at Howell like a club.

Howell caught the man's wrist, and pulled him close. They were eye to eye. Howell was ready for a fight, but the other had planned on a very different kind entertainment this evening. "Bugger off!" Howell shoved the man away.

There would be no sausages for Howell. He passed the first booth in the tent city where a buxom lass was handing out foamy mugs. There would be no smiling female for Howell, either, nor even beer. Music and laughter, and the smell of meats and cheese and roasting nuts penetrated the canvas booths around him. A woman was singing. It had been just the same when he was a boy walking down the grandest streets of London, his stomach growling and his clothes in tatters.

He passed the moneylender who they had seen vex a man earlier. The man who had made the loan was in earnest conversation with another reluctant customer. The injustice of it chewed on Howell.

"Davidson! Get your arse over here."

Howell turned to see Ruzeque, the French wine seller. He had a few customers drinking his wine, but was certainly not mobbed.

"And who might this be?" asked Giovanni.

Howell turned and began to walk towards Ruzeque. "Do you recall I told you that after we fought the pirates, Rosy slipped out and she and I sold her tobacco? We did it front of Ruzeque's wagon."

Ruzeque looked hard at Howell. "So what are you going to do to save my damn business this time?"

Neither Howell nor Giovanni answered.

"No gorgeous red-head selling smoky bullshitte to attract the marks, eh?"

Howell was sorry that he had to admit the truth. "I am afraid not."

"So You are going to buy wine, then." Ruzeque looked at them as though demanding that they patronize his establishment.

Howell and Giovanni looked at one another shaking their heads. "No." Giovanni held up his purse and shook it like a bell so there would be no doubt that it was empty.

"As God is my witness, sometimes you make me feel like a simpleton." Ruzeque pulled a mug of wine for each of them and handed them to the mates. "Here."

"Thank you, Sir."

"Molti grazie."

*Yah, yah* "I am sure you'll pay me back one day." The sarcasm was so thick you could have used it as mortar to set stones.

Howell began to feel his anger subside. The wine helped, sliding down his throat and warming his stomach, but equally it was the fact that someone had cared about him. Ruzeque had a peculiar way about him, but he had helped Howell when he needed it. Howell had not experienced a lot of that.

"So, what do you two penniless babes got in mind for the evening?"

Giovanni considered this. "Well, first I am going to look at the girls selling bread and cheese. Looking is free, you know. Then I am going to look at the girls who are selling themselves. By then I will be exhausted, so I will sit and…"

"Fuck that. I am sure I will get some pretty rowdy drunks tonight, from those who *do* have money, you understand. I could use someone who stays reasonably sober and can kick arses as required. You want the job?"

Giovanni gave a Howell a questioning look.

Howell swept his hand towards Ruzeque. "Do as you wish."

Giovanni did not move. He looked at Howell. "What will you do?"

"I have an idea to find Rosy."

"That red-headed stink seller?" Ruzeque just shook his head. "Good luck with that."

"I do not think that luck will be required."

"Really." Ruzeque was looking at Howell with genuine interest.

"They do not have a permanent place in town, so I will look for the biggest house in the nicest neighborhood. They probably have that blue and white flag outside, as well. Good night gentlemen, and good luck, Ruzeque." Howell pulled his battered hat low over his eyes, directed his worn shoes back towards the city, and away from all the unobtainable temptations in the riotous festival behind him. He returned through the great gate that the Regiment had used to leave the city, and began his search. He traversed the warren of Nurnberg's streets, passing churches, stern-looking clergymen, preoccupied businessmen, lodging houses, packs of street urchins, smoke-filled workshops with smiths beating red hot iron with their great hammers, commercial properties, and the constant flow of wagons and porters and women carrying baskets.

That vermin Sir Horace seen and lost! His pay dangled before him and snatched away. Blocked from even something so basic as food. If he could again be with the breath-taking Rosy, it would all be worth it.

He followed his nose into more and more affluent parts of the city. At length, his feet brought him to an unusually wide street. On the one side was a row of shops, but these weren't the kind that would welcome men of his sort. The signs did not feature the obvious drawings that attracted the illiterate masses, but were elaborately lettered for those who could read, and who could afford the quality goods and services that were on offer within. There was a lawyer, and a jeweler, and several doctors.

In a few more steps he saw a great house flying the blue and white flag he had seen during the review in the main square. Hullo! This must be where the Countess Elizabeth is staying. So this is where the delectable Rosy must be, too. Could he manage a meeting with Rosy? For preference, one that lasted longer than two words exchanged on a windy mountain pass. Howell was confident that the direct approach would be unsuccessful. '*Good afternoon Mr. Butler, Sir. I am a grubby, penniless soldier. Is the lady of the house receiving?*' Not bloody likely. Not for his kind. Not for the 'lower classes'. Order and propriety and knowing one's place rose up and slapped him across the jaws.

*Oow! Oow! Oow!* His unseemly call echoed off the pompous businesses on the one side and the imposing mansion opposite. Howell's noise was totally out of place, which was exactly the way Howell felt.

He was never going to draw her out by barking in the street, however. He needed a less obvious way to contact her, and arrange a meeting. He slowly walked across the front of the great house to a small, elegant park on the side. Beyond that he found an alley leading down past the greenery to the back of the house. The flowers, trimmed bushes, and formal trees in the park did a tolerably good job of obscuring what lay within, so Howell decided to take a little walk down the alley. As he approached, he saw a barn on the side of the alley across from the mansion. The working part of the garden was next to the grand residence, shielded from the street by the manicured greenery. Next to a small gate was the cesspool, and further along there were rows of vegetables growing. From the noise coming from this end of the house, it sounded like the servants were preparing the evening meal.

Howell slipped through the gate, found a secluded place to lounge, and settled in for a bit of observation. Presently a girl emerged from the kitchen door, and spent some time amongst the vegetables. She was earning her keep. She was preparing food that someone else would eat. People who thought so much of themselves that they could not be bothered to cook their own meals. Suddenly, irrationally, he saw the poor wench in the garden as an accomplice in the injustice he saw all around him. If she and her kind refused to do for these self-centered worthies, why then they had have to get their fingers dirty like the rest of us, wouldn't they? *oow! oow! oow!* he chastised her, this time not so loudly as before. It wouldn't do to reveal his new position to whoever might be inside. The girl leapt up in confusion and alarm, and hurried into the house with her arms spilling their leafy burden.

Howell heard a commotion within, the kitchen door crashed open, and two females hurriedly emerged. For an instant Howell could not separate them, but then he got a glimpse of a red head that looked very familiar. Rosy conducted the vegetable-gatherer back into the garden, and said something that Howell could not make out. The poor girl gestured in what was vaguely Howell's direction, and said something in reply to Rosy.

Rosy approached his position, and scanned the trees and bushes. "Well, *I* do not see any dogs or werewolves or whatever it was that you thought you heard." She unknowingly brought her delectable self a few steps closer to Howell's hiding place, still searching. "Quickly now! Finish your work."

Howell continued to sit quietly, content to remain unseen, and to take in the show by peeping through the leaves that surrounded him. Rosy's curves and movements cast the same spell over him as they had at the city square. Her black and grey attire had not seemed special when compared to that of the Countess when he had first seen it, but it was certainly elegant compared to the patched suite worn by the kitchen maid. Howell quaffed Rosy in deep drafts.

The scullion was completely absorbed by her harvesting, and Rosy was about to turn back to supervise her when Howell stood. He said nothing. He made no sound of any kind. He did not even try to attract Rosy's attention by waving his hand. It was enough. She immediately focused upon him. She smiled. Ghostly, Howell nodded, sat back down, and disappeared.

Rosy composed herself. Still, however, she retained the merest hint of a smile upon her face. "Have you finished yet?" Rosy did not take her eyes from where Howell had vanished.

"Not quite Miss. I still must fetch some of the leeks you see before you."

"Deal with them later. Take what you have and go inside."

"Yes, Miss."

Rosy waited for the girl to leave the garden. She watched carefully out of the corner of her eye until at last the kitchen door closed behind her. They were alone.

Howell stepped from the shadows and advanced upon her confidently. He whispered, *oow, oow, oow*

Rosy rushed into his arms. "Howell, Howell, it is so wonderful to know that you are safe, and now you have come to me." She hugged him tightly and then took him by the hand. "Follow me." She led him out the little garden gate by which Howell had entered, across the alley, and into the barn. "We have important things to discuss."

"Indeed we do." The stable had the warm smell of well-tended horse flesh, and a positively spectacular carriage in which the Countess was accustomed to ride. Howell took her to the coach. "Let us make ourselves comfortable, shall we?" He held open the door, and ushered her in.

He followed her in, and she immediately fell upon him. She wrapped her arms around his neck and kissed like only reunited lovers can kiss. She pulled back for just an instant, to look at him and to perhaps reassure herself that he was really here, and then she kissed him again. Once again she released him, and she slid down to place her head upon his chest. Her emotions passed from the frenzy of knowing that he was safe, to the passion of knowing that she was with him. His chest rose and fell beneath her head. "I saw you fighting your way through the crowd at the parade, and on the mountain you said that Sir Horace had betrayed you." She sat up to give him a firm look. "What on earth happened?"

It was wonderful to have found her again, but she was right that they had pressing matters to discuss. He released her, and sat back. "Someone fired a pistol next to me, and immediately took off running. I tried to catch him, but the crowd panicked, and starting running every which way. When I tried to force my way through, it looked like I had fired the shot."

"Who was it that had fired?"

"I am not sure, but I imagine that it was one of Horace's men. I ran and dodged the guards for quite some time before I got to your residence. While I did not think that Sir Horace was my friend, I thought that at least he would want to protect someone that you, well…"

"Loved."

They looked deeply at one another. Howell supposed that what he felt must be love as well, but he just could not know.

Rosy had expected Howell to respond in kind, to express his love for her, and when that did not come, she swiftly returned to Howell's story. "So, what happened when you found Sir Horace?"

"He was in the front hall. Looking back on it now, he was just waiting for me so that he could lead me into his trap. We went out onto the street, one of his roughnecks grabbed me, and he pulled his rapier. Then he called for the guards.

"Good God."

"I barely dodged his thrust, but still he cut my side."

Rosy went silent as she took this in.

"They threw me into a stinking dungeon that was like a pit in Hell."

"Oh, Howell." The pain she felt covered her face.

Howell felt the stab of the betrayal once again. "Sir Whore's Ass will pay for this one day."

Rosy smiled at the ironic name that Howell had given to his adversary. "Howell, darling, if you were in prison, how do you come be here?"

"I had the help of another prisoner, and we managed to break free. Horace and the castle guards chased me throughout the city for a second time in as many days. Running where I might, dodging down alleys, disguising myself. He shot at me as I finally left the city. My friend and I joined Mansfeld's Regiment to have a place to hide."

Rosy bit her lip, holding Howell's hands. "So that is why you never returned to me that day."

"Believe me, it was not of my choice."

"What will you do?"

"For the time being, I shall just keep soldiering, and keeping my head down. I made a fast friend in Giovanni, the man who helped me escape the cells." Howell smiled at Rosy. "Fortunately, now You are here with your chum Elizabeth again, so Horace's plan will come to naught."

"Alas, that is not true. I overheard a conversation between the Count and Sir Horace recently. He claimed that he was a secret agent sent by King James."

"But that's rubbish," said Howell. "King James is determined to have nothing to do with a war in Germany."

"This is true, yet Sir Horace presented all the musketeer kit to Fredrick as a present that supposedly showed that James has changed his mind."

"And all the muskets and lead and whatnot are useless because there's no powder," Howell said.

"Yet Horace is quickly establishing himself as Fredrick's chief advisor."

"God's blessed mother!" Howell fairly jumped from the seat. "You are at the mercy of that monster?"

"Howell, please calm yourself."

"Calm? How can you ask me to be calm when at any moment Sir Horace could put a rapier through your heart?"

"No," Rosy replied calmly. "No, he cannot."

"What?"

"Count Fredrick is fascinated by Sir Horace, but Elizabeth is not so easily persuaded. Elizabeth has been urging Fredrick to caution regarding Sir Horace. She has spoken to her husband several times when I was in the room with them, so it would be plain that I was privy to her thoughts. If Sir Horace were to kill me, Elizabeth's closest friend and confidant, then Count Fredrick would see it as proof that Elizabeth is correct, and that Sir Horace is in fact a threat to them. Sir Horace cannot touch me without making his position worse."

Howell grinned. "So I may satisfy my thirst for revenge at my leisure."

"At your first opportunity."

"Now seems an excellent time."

"He's not in Nurnberg, Howell. Elizabeth is here to greet your Regiment, but Fredrick and Sir…" she paused with a mischievous smile "…and Sir Whore's Ass are in Heidelberg." She became serious. "I cannot allow him to lead Fredrick and Elizabeth over a cliff. I want to kill him, and be done with his efforts to control Count Fredrick."

"Killing is my job, not yours."

She ignored him. "Can you smuggle a pistol to me?"

"How would you explain the noise when you fired it?"

"Then I will use a knife. There is certainly something I could use that's in the kitchen."

"You have no experience, so you will not make a clean job of it. He will doubtless struggle, and make noise. They will find you with stained clothes and a bloody knife in your hand. Whether you succeed or fail, you will be seen as an enemy of the revolution, and Elizabeth will no longer be able to restrain the Count. You'll cause the Revolution to fall into the Emperor's hands."

She showed her frustration. "Then you must do it."

"As soon as I can."

He put his fingers onto the back of her neck and, ever so slowly, pulled her face to his. There was desire in her eyes. He pulled her head towards him and her lips onto his. They kissed long and deeply.

He was not sure how he felt about this whole situation. He hugged Rosy, and thought. He desperately wanted to see the hated Sir Horace served his bloody portion of retribution, but how he was to get close enough to kill the man was quite the mystery. The one thing of which he was certain was that he wanted this woman. He wanted her body. He wanted to know how she felt. He wanted what she said to him, and it was not just what was happening in the Palace. He wanted to hear her talk about he and she as one.

Suddenly she pulled back, her eyes riveted on his. "I must go back or they will miss me. We cannot risk anyone knowing that you are here."

# Chapter XIII

## Storming Pilsen

The fog drifting across the disk of the rising Sun gave it the appearance of a full Moon soaked in blood. Nor had the coming of the dawn infused any warmth into the icy mist, and the men had been standing in formation for an hour or more. Rank after rank, the musketeers disappeared into the fog ahead and behind Howell and his mates. All around him, the soldiers flapped their arms against the rough wool of their doublets, stamped their feet, and complained to one another with ragged puffs of steam coming from their mouths.

"Porca miseria!" Giovanni rubbed his hands together vigorously. "Cannot we just forget all about this hog wallow?"

The question foremost in Howell's mind was very different, however. He was anxious for the battle to begin. He had been in street fights with perhaps a dozen boys all told, and the fight with the pirates had involved perhaps a score or two on each side, but the coming battle would involve hundreds of men, perhaps a thousand. It was yet another milestone on his march to becoming the consummate warrior. He looked about at the faces that surrounded him. He expected that many of them would soon be wounded. Some of them would be killed. Howell felt only a weird species of exhilaration. It would be a deadly dangerous business, but he could not suppress a morbid thrill at the prospect. Something deep below his conscious thoughts was sniffing blood.

Giovanni put his hands into his armpits. "I just hope this is over with quickly."

Howell knew that this wish could be fulfilled in a number ways, including Giovanni's wounding, or even his death. "Be careful of what you wish for." This would be Giovanni's first battle, and it would be one Hell of an introduction to what it meant to be a soldier. How would Giovanni react?

Giovanni just stared at Howell, confused by what his friend might mean.

*Ha!* An old ranker's toothless face beamed. "I am wishing for gold!"

"I am wishing for a woman," said one of the other soldiers.

"Lots of women!"

"Lots of gold!"

"A barrel of beer just for me!"

Their quiet little comrade got Howell's attention. "Not woman I want. Not beer. I want why us here?"

*Bang!* Their little cannon fired. The ball went whizzing through the fog towards Pilsen. It was one of the guns that had battered the hole through which they were about to enter the city.

"There's a good question. This is Bohemia, no?"

*Hmm?* Howell's thoughts were wrenched from his battle preoccupation.

"We are in Bohemia, are we not?"

Giovanni had not lost his talent for abruptly re-directing Howell's attention. The interruption was not welcome. Howell scowled. "Surely you recall that long walk we took some weeks past through parched fields and dripping forests and mountains haunted by wolves?"

Giovanni smiled. "That lark?"

"Yes, that lark. A journey like that is how one gets from Nurnberg to Bohemia."

*Uh hunh*

"And how one wears out another pair of shoes!" added the ancient warrior.

"Fine," persisted Giovanni. "Which brings us to this land of fogs and frustration at the invitation of the Bohemian Es… Es…"

"Estates." Howell was impatient with all this. "The Bohemian Estates. It is something like their Parliament."

"How nice for them." Giovanni blew on his hands. "The point I am trying to make is that we were asked to come to Bohemia by these Bohemians, and here we are attacking a Bohemian city."

"Why us here?" The young soldier was plainly frustrated.

Howell sought to end this distraction. "Most everyone else in Bohemia has told the Emperor to bugger off, but these fellows are of the opposite opinion." This was not the topic that Howell had been dwelling upon and, frankly, the 'why' of his first test in a real battle did not interest him.

"Yes, yes, and Budweis is holding out too. We know that. But what we want to know is why?"

"Just like you, there's always someone contrary in every group."

Giovanni threw up his hands in frustration.

*Boom! Bang!* The distinctive sounds of their small battery of cannon shook the air. They had spent weeks outside Pilsen's walls, waiting for Mansfeld to scare up some artillery, unable to do more than shake their fists at the defenders. The two guns that the Colonel had finally managed to hire were on a little rise of ground close enough to the walls that they could do the required damage. One was substantial cannon that had taken oxen and swarms of men to emplace, but the other was so small that it was flattery to refer to it as a cannon at all.

Sergeant Blutworth now emerged from the gloom. "Right! Gather round." He took a stand in the midst of his musketeers, and his men shuffled into a rough circle around him. He had looped across his shoulders several pieces of match cord, each about a fathom long, and smoldering at both ends. Blutworth handed them to Howell. "Hand these out, Corporal."

Howell realized that he was not just responsible for himself. He had a Squad of men to lead. He had mocked Vlok for not thinking of his men, and now he had to live up to the standard he had set. He wouldn't be ordering men to push wagons this time, he had be ordering them to go into battle.

Without the coal on the end of a burning match cord to ignite the powder, a musket was useless, but since match cord cost money, only a small minority of musketeers would normally have theirs lit until battle was just about upon them. Howell passed down the line, giving some cord to every eighth man so that the musketeers could light their matches when the time came. The designated men kept most of the tangle of cord stuffed into his bandolier and the smoldering ends between his fingers.

The soldiers attending Blutworth were very young for the most part, hardly older than boys. No horseplay today, though. At any other morning stand-to they had be laughing and shoving one another and the very Devil to keep in order. Today, they stood quietly in the fog, every one of them clutching the muzzle of his musket with his eyes upon his Sergeant. Their faces ran the gamut from eager to mystified to glazed by drink to contorted with fear. The mouth of the man beside Howell was slack. Behind him another was noisily sick, puking into the mud at his feet.

*Bang!* Their smaller gun fired off into the breach, echoing through the fog, and giving a menacing air to what was going on.

"What lies before us is a test." Blutworth looked around at the faces fixed upon him. "A test of our courage, yes, but more, much more. This will be a test of our comradeship."

This seemed to catch many of the men by surprise. They had expected the shooting and the noise, and the more realistic men must have acknowledged that there would be blood, but they had not realized how important their Sergeant regarded brotherhood. He had impressed it upon them on the mountain side, and now he had returned to this topic.

"You will be pressed from every side with the Devil's own racket. Men will be yelling all 'round you. There will be musketry, and there will be the cries of the wounded. If the man beside you gives way and runs, what will you do?"

The lads considered this. They looked at one another. On and off, the thought had gone through their minds all the way from the Alps. Now they were face to face with it.

"If you give way and run, what will happen to your mate?" Once again, Blutworth gave them a moment to chew this over. "We are in this together, lads. It is not about Mansfeld's Regiment, it is not even about our Company or Davidson's Squad. It is about the men you see about you right now. We will be scattered like so many sheep, or we will stand together and taste victory together. The choice is yours."

Howell could see a noticeable improvement in the confidence of the men. Not a few times did he see men shaking hands. Many hugged. These men were coming together. The immediate future would be a defining moment in their lives. Almost all of them had come from the kind of backbreakingly poor, despised, and powerless origins that Howell had. Now they could see the prospect of gaining their first real strength, and their first measure of respect.

"Have them load, Corporal." The Sergeant had others to encourage, as well, so he disappeared once more into the fog.

"Our Sergeant has ordered us to load our muskets." Despite Howell's promotion, Corporal Davidson preferred to arm himself with the musket his men carried.

Even that minority of Mansfeld's Regiment who had seen action before had not performed the complex task of readying their muskets since last Autumn. Howell walked amongst his men, checking their work, and performing his first time as a Corporal. Vlok glowered at Howell doing a job that he believed should rightfully have been his.

For Howell, it was a ceremony of preparation. Mr. Grace had told him the great story that Homer had written twenty five hundred years before. Howell felt like Achilles purifying himself before battle. Howell was not just a musketeer, but a warrior as well. He was carrying on a tradition that stretched to the limits of the memories of men.

Loading had many steps, and took a long time. Even Howell, for all his eagerness, could only speed things along so much. Giovanni was going through the same process at a somewhat more leisurely pace. "You know, if we can draw this out just a shade longer, there may be time enough for the defenders to freeze to death."

*Boom! Bang!* Two more shot flew off towards Pilsen's defenses. Were they really trying to hit anything, or just firing blindly into the fog?

From the front of the column came another sound. This was of a great many men speaking and shouting. The clamor grew louder as it grew closer, and soon a party of horsemen materialized, their mounts walking alongside the column. They were led by the Regiment's intense little Colonel, Ernst von Mansfeld himself.

"Gott mit uns!" he cried out. "Today's password is 'God with us'!"

Giovanni scoffed. "I do not know about God, but at least we've got that big-balled horse of his with us."

"Hey, Colonel." A brandy-emboldened musketeer called from within the column. "When are we going to get this party started?"

"The party comes later." Mansfeld called back happily to the soldier. "First we show these shopkeepers and dung-haulers what it means to face real soldiers. Gott mit uns!"

"God with us!" "Let's do it!" *Hurrah!*

Howell did not care if his eagerness to do battle with his fellows was God's blessing or the Devil's curse, he simply yearned for the fight.

*Bang!*

As the cavalcade disappeared towards the rear of the column, Sergeant Blutworth walked down the length of the column, restoring order amongst his men. "Quiet! Settle down. Right. Get your matches lit."

The men clustered around the few who had matches that were smoldering.

"I have it," said Giovanni. "We will surprise them. We will creep up on them in the fog, and stick these coals up their asses."

"Form up! Form up!" Blutworth called.

Howell lined up his files, and straightened his rows. The Regiment's shotte were being assembled into a heavy column six files wide and many score ranks deep.

The Regiment's spears, however, were nowhere to be seen. Those eighteen-foot poles were barely manageable in the open field. In the tight confines of the breach, they would be worse than useless. If Colonel Mansfeld was going to proclaim himself master of the city of Pilsen, it would be due to the efforts of his musketeers.

The walls of this ancient town had been constructed uncountable generations ago. Against battering rams and archers they must have been formidable, but faced with modern artillery, tall towers and walls only gave the illusion of security. In a surprisingly few days of bombardment, these defenses that had stood for centuries, and defied Mansfeld's Regiment for weeks, had yielded a hole big enough to drive wagons through. Howell had watched the process intently, and knew that on the far side of the ragged notch that had been hammered through the masonry there were seemingly undamaged buildings, frowning down upon the breach. He and his comrades would not be going through the opening in the wall and into an empty field, but rather emerging into a sort of amphitheater of combat.

*Boom! Bang!*

"Prepare to advance!" The command echoed through the fog from the front of the column, repeated by every officer who heard it. "Prepare to advance!"

Howell eagerly snatched up his heavy musket. He was buzzing. He wanted to do this. He needed to do this. Without conscious thought, he pulled out a little more matchcord from the tangle at his hip. He adjusted the smoldering match ends between the fingers of his left hand to prevent them from singeing him. He was in the grip of a terrible desire. He saw in the breach ahead of him the fulfillment of his most fervent nightmare.

Beside him, Giovanni observed Howell's wild eyes with concern. "Are you well, amico?"

Howell could not suppress a grin.

Giovanni's appearance, on the other hand, became even more grave than it had been before.

"Shoulder arms!" A thousand muskets were hefted onto a thousand shoulders. The musketeers would carry their weapons as so much dead weight until they were in the face of the enemy.

*Bang!*

"Forwards!" The exhaust from the men's lungs mingled with the lingering fog. The whole clammy brew was scented by nervous sweat, vomit-flecked doublets, and the sulfurous odor of match smoke. Thousands of wooden chargers swung on the cords that suspended them from their bandoliers, and rattled against their neighbors.

*Boom! Bang!* An instant later the iron spheres zoomed by. Howell heard a loud crash as the larger ball slammed into something solid ahead of him and the sharper sound of the smaller shot slapping off some bit of masonry. These would have to be the last shots from their guns, as the head of the column would surely be in the breach momentarily. Howell was so captivated that it was all he could do to keep from running.

*Crack! Snap! Crack! Crack!* Uncoordinated musket shots rang out of the fog. The defenders' fire was premature and aimless. These were surely amateurs they faced. A musket ball hummed by just overhead, the sound like what he had heard when Vlok had discharged his musket over their heads while they were training. It reminded him, too, of the zip of the pistol ball that Howell had dodged at Genoa's gate. Someday Sir Whore's Ass would regret having fired that shot.

"Ha!" Giovanni shook his fist. "You are going to have to do better than that!"

*Snap! Crack!*

*Ahh!* Towards the front of the column someone yelped, and their formation was disrupted.

"Forwards! Keep moving!" Corporals shouted to keep the men from stopping. The files wavered like streams of water diverted by a boulder. Howell tried to follow the man ahead of him as he veered, but suddenly two musketeers were standing motionless directly in front of him. One of them was supporting the other, and the latter had fresh, bright blood all over one side of his weathered, grey doublet. He looked stunned. The helper hustled his comrade out from Howell's path, and onto the wet field. This was real. Howell had made the ultimate choice, and now he would enjoy, or suffer, what followed.

"Move! Move!" The Corporals grabbed and shouted to keep a semblance of formation and progress. "Close up! Move! Move!"

Howell trod on something hard and realized it was the musket dropped by the wounded man. Three hundred kreuzers cast aside like a stick in the forest. A man chopped up like an animal for slaughter. Would Howell be wounded? Killed? It just did not seem possible. He felt indestructible. The notion filled him with a sense of defiance. *Oow! Oow! Oow!*

*Hurrah!* A multitude of shouts split the fog. "God with us!"

The sound of the musketeers' advance now included the clattering of shoes on loose masonry fragments. Howell could see the heads of his comrades at the front of the formation beginning to rise like phantoms as they climbed up the rubble. Then, immediately before him, he saw a broad apron of shattered bricks and stones. These were the remains of the once mighty wall and the two structures that had backed up against it. All of this rubble had settled into a sort of crude ramp that was strangely white, as though sprinkled with snow, but the color came from the plaster and mortar pounded out of the buildings and walls. Rank by rank, grunting with their exertions and shouting encouragement to one another, the musketeers stumbled and crawled up this slope and into Pilsen.

Howell and his Squad passed where the defensive wall had once been. Howell could see the ragged edges of the wall, battered through by the fire of their guns. As he entered the city, and started down the broken bricks on the other side, he saw the interior of a three-story building that appeared to have been crudely sawn in half from the top to the bottom. The house had been made using the city's outer wall as its back, so that when the bombardment had destroyed the defenses, it had destroyed this house too, or half of it leastways. He could see a bench still resting in one of the half-rooms, and a picture on the wall above it. These homey artifacts clashed with the sagging, splintered timbers that were all that remained of the floor, and with the raw bricks exposed in the walls. There was a similar half-wrecked building on the other side of the breach as well.

Ahead of them was a wall of mostly undamaged buildings, or at any rate these had only the round holes where a shot had punctured them. The inn directly ahead of Howell had at one time drawn customers with a large painting of a rooster. There were other advertisements on some of the buildings and some appeared to be private homes. Their advance was very effectively blocked by these various structures.

The leading ranks descended the uncertain footing on the far side of the breach to gather in the confined space amongst the buildings. They were not advancing, though, but merely finding places for themselves in the crowd. The Corporals had no idea of what deployment the men should take, so the troops remained a formless mass. No one seemed to know what he was supposed to do.

Howell was almost frantic to maintain his progress. Why did not they move? What was the delay? He pushed men aside to make a path through the press, looking for a place where he could find a target and deploy his men. He could barely see through the fog and the drifting match smoke and the milling heads of his comrades. He caught brief glimpses of enemy musketeers posted in the windows of the pockmarked buildings around him, as well as at some of the larger holes made by the cannon balls. He saw an enemy musketeer posted in a doorway on the top floor of one of the half-wrecked houses. Pilsen's defenders looked down upon Mansfeld's Regiment as though they were observing animals in a pen.

"Prepare to fire!" Blutworth and the other officers knew what was coming even if their simple subordinates did not. They forced their way frantically amongst their clueless soldiers. They were trying to organize lads that had never done anything more violent in their lives than exchange angry words with a vendor at the weekly market. "Prepare to fire!"

The loose rubble provided terrible footing, but the soldiers were too tightly packed to deploy their weapons properly anyway. Some of their men fumbled with their clumsy muskets trying to follow their orders. Some were just engulfed in the chaos around them. Some of the veterans in the ranks realized what was about to happen, and more than a few of them froze in terror.

Howell, too, sensed that something awesome was coming.

A trumpet blared, cutting through the mob of Howell's mates trying to form. The Mansfelders froze at this ominous sound from the enemy. An eerie silence pervaded the breach, broken only by the sound of a man sobbing. Howell saw the enemy musketeer on the top floor of the sawn-through house blow his match into a bright coal, and force it into the end of his cock. He aimed it at the mass of men below him. The trumpet sounded a second time.

Every Pilsener musket surrounding the breach discharged at once. A terrible roar and billowing smoke filled the air. Musket balls buzzed in from all directions. Every shot in the mad swarm found flesh. Smoky, shouting, screaming chaos engulfed the Mansfelders. Some were dazed and uncomprehending. Some froze. Some attempted to force their way past Howell and escape the seething, bleeding mass.

Something inside Howell's mind went incandescent. His focus narrowed to a point. There was the here and the now, nothing else. Just ahead was something beckoning him that had haunted him all his life. He put his head down and began to force his way through the riot before him.

"Amico! Where are you going?"

Howell did not know what lay ahead but he knew that he wanted it with all his being.

The wounded were everywhere. Their screams filled the air. A man staggered into him, his face covered in blood. Howell worked his way past another who clutched one of his hands in the other with blood dripping from them both. Howell wanted something, needed something. Whatever it was, it lay somewhere just ahead. He forced his way through the crowd and found himself at the top of a small hill of rubble. Here he could see over the heads of his smaller comrades, but he saw nothing. No enemy. No threat. No prey. Between the fog and the powder smoke and the defenders' natural desire to conceal themselves while they reloaded, it seemed as though the only people present were the boiling Mansfelders. They were shouting and crying and scrabbling all around him. Howell, though, felt like the lone champion.

He picked out a break in the encircling defenses where a half obstructed street left the battlefield. Between the rooster-marked building on one side and the sawn-through structure on the other, the defenders had sought to seal off this exit by constructing a barricade of rubble across the road. Howell could see the heads of enemy musketeers lining this crude wall. Their right arms pumped up and down as they worked their ramrods. They were only a few dozen paces away.

Howell warmed his match and cocked it, and then made a conscious effort to calm himself. He held his breath and aimed his musket at a head that projected above the parapet and was covered with a knit cap and a full beard. Howell squeezed his trigger and pulled the cock slowly back until the glowing ember touched the powder in his pan. *Flash! BANG!* The heavy musket kicked back into his shoulder. The scene before him disappeared in the smoke from his weapon's discharge.

Howell was shocked as a volley burst out around him at almost the same moment as he fired. It was his squad. They had forced their way through the maelstrom in his wake, and now were following his example.

These were the only shots that had been fired since the terrible volley by the Pilseners, and the report of these muskets echoed between the scarred buildings. Some of Howell's comrades cringed, it is true, thinking that they were being fired upon once more, but most of them realized that the shots had come from their side, and that it had not been answered.

Howell held his musket over his head. "Comrades! Brothers! Join with me! Fight with me! Victory!" He felt unconquerable. *Oow! Oow! Oow!*

That one volley and Howell's call to action were tonics for his terrified mates. Faces that had been filled with confusion and terror only a moment before, now looked at Howell with hope and resolve. Mansfeld's Regiment had fallen into the Pilseners' trap, and that first volley had been horribly effective, but it might yet prove a mistake. Every defender's musket was now empty, and it would take some time to ready them for another shot.

"Reload," said Howell.

The other Corporals moved amongst their men, giving direction and restoring order. The mob calmed. "Face front. Cock your matches." They heard their orders. They tended to their matches. The Pilseners' volley had caught them before most of them could fire, so, one by one and then in small groups, the Mansfelders found something to aim at, and loosed their balls at their enemies. The next fellow shuffled forwards, and found his target. The Mansfelders had regained their balance. It would be a battle after all, and not a slaughter.

Howell feverishly shook the contents of his second charger into his musket's muzzle and fumbled a ball out of his shot bag. A Pilsener projectile buzzed by thrillingly close. He worked his ramrod out of its storage slot, almost snapping the stick in his eagerness, turned it end for end, and dropped it down the barrel. *Crack! Crack!* Shots rang out at irregular intervals on every side. Some of his comrades were shuffling into position on either side of him and, behind him, some were shouting defiance. He tamped down the charge in his musket with a few strokes of his ramrod. *Crack!* Ahead of him, he heard commands shouted in some language he did not know. Czech. It must be Czech. He pulled his ramrod out of the barrel, flipped end for end, and restored it to its storage space under the barrel. *Crack!* Bullets just kept coming. *Crack!* He had not known such exhilaration since his escape from Genoa.

This barricade was the key. Once they had passed this obstacle they would be in the city proper, instead of being trapped in this shambles. The street blocked by the barricade was perhaps twenty feet wide, and the great dike of rubble was topped with the heads of a dozen or more defenders. Howell saw these men place their muskets on the top of the barrier, aimed right at him. Little curls of smoke arose from each of the dozen matches. There was a curt shout from behind the defenders. All the dozen muskets spoke at once. The top of the wall was obscured in the blooming grey-white smoke. Howell heard the crackling of their discharge and, and instant later, he felt a burning sensation in his calf. He had been hit! He smiled. He almost laughed. His enemies were doing their best, and all they could do was nick him.

The smoke that obscured the enemy position began to clear and he saw one row of heads replaced by another. "There!" he called to the men around him. "That barricade. That's the key. Shoot them down." His squad had found places beside him, pointing their muskets at the wall. The enthusiasm of some could be mistaken for madness. For once in his life, Giovanni was somber.

Their foes had their muskets on the top of the wall again.

Howell took his aim as best he could in the shapeless rubble he stood upon. "Ready!" He blew his match to give a strong, glowing coal. He slid open his pan cover. "Give fire," he called out. Most of the Mansfelders around him fired at Howell's command. The musketry around him sounded like an unstoppable power. The smoke around him smelt like an irresistible force.

Another volley roared out from the enemy position. A wounded man screamed into Howell, and knocked him down the pile of splintered timbers and chunks of brick. Howell clung grimly to his musket as he rolled painfully down the jagged shards. He tried to protect his head, but had to leave the rest of his body to fend for itself. He rattled and bounced down the slope, flew through the air for just an instant, and landed with an impact that nearly drove the wind from his lungs. He lay there in the little cloud of dust he had raised. Tangled in his equipment and sore from a hundred minor blows, he swiftly regained his senses.

He also regained his craving for battle. He stood. He twisted his bandolier into position. He found that he had fallen into the ground floor of one of the structures that had been sawn through from top to bottom. This was the one next to the barricade that Howell had taken as his target. Less than half the wall remained and the joists above him sagged down. He freed his chargers, most of whom were tangled with others. The building's door that had led out into the street was missing, replaced by a cascade of rubble. Howell adjusted his shot bag and his length of match. He was a musketeer again. His first item of business then was to reload.

As he worked his way through the procedure, though, he looked again at the street door and its filling of broken masonry. It occurred to him that this must mark the location of the barricade that blocked the way just outside. He quickly surveyed the rest of the room. Everything was thickly covered in plaster dust. Across the width of the house, very nearly opposite where he stood, was a door that looked like it might communicate with the building next along the street.

A thrilling fancy gripped him. It felt like a vision of the future. He quickly made his way to the door in the far wall, cracked it open, and peeked through. Like the room he occupied, this space was empty. It had another wall, this one intact, and in this wall was a door that led onto the street.

Howell crept into the second room, and barely cracked open its door. Sure enough, he could make out enemy soldiers hurrying to and from their barricade. He was behind the defensive wall. He could hardly contain his joy as he quickly retraced his steps. He had found a way to turn the enemy's position, and attack them from the rear! It was a revelation to him that he could not only be the champion and warrior as he had always seen himself, but that he had a talent that could clearly show him the tactical realities of the situation as well.

He had found a great opportunity, but he had need more than just his own two hands to use it. He regained his position in the first room looking over the half wall, and up at the battlefield. The ragged slope extended from where Howell's position up to where hundreds of his comrades stood. There were no signs of panic now amongst them, just the little of the cries of abuse or pain that he had heard before. A man aimed his musket, and took his shot in a thoroughly workman-like manner. Another found his place as the first retired. A wounded man was helped to the rear. "Keep moving!"

*Oow! Oow! Oow!* God's wounds! How was he going to collect some men and execute his plan? Then he saw Giovanni. He was in the process of taking his place in the front of the pack. He looked around, trying to locate the source of his friend's barking, but without success.

Howell needed men, so there was nothing for it. Howell quickly scrambled up over what was left of the room's wall, and stood so that he was completely exposed. *Oow! Oow! Oow!* He gestured broadly, and got Giovanni's attention. He also attracted the attention of some of the locals, as well, and one bullet, and then another, hummed by him. *'Come on! Come on!'* he beckoned with his hands.

Giovanni came scrambling down. He was followed the rest of Howell's squad.

Just behind them Howell saw Vlok. He glanced at Howell a few times, seeing what was going on, but mostly he was pointing men at the barricade, keeping himself well back, and making sure that the enemy did not see him.

Then an unexpected face appeared before Howell. It was Sergeant Blutworth, and he demanded, "What's all this then?"

"Begging the Sergeant's pardon, but I have found a way to attack the enemy from the rear."

"Do what?"

"There is a door that leads into the next building just there." Howell pointed over his shoulder with his thumb towards the wall behind him. "And from there is a door that leads out onto the street behind the barricade."

"The Hell you say!"

"I swear, Sergeant, it is…"

"Right!" Blutworth shouted at Howell. "Count slowly to five hundred and then make your attack!" He could be heard shouting "Cease fire! Cease fire!" as he hobbled up the slope.

Howell turned away from the Sergeant. The clarity that he had felt with the thrill of battle was now transferred to his new goal. His focus was awesome in its intensity. His hands and his mouth unconsciously tended to his match while his mind evaluated taking his first command into a planned attack. He was going to be very busy getting this all together. "Giovanni, count slowly to five hundred."

"Uno, due…"

"Is everyone loaded?" Howell asked. Everyone nodded except one, but he was hard at it. Howell primed his own weapon.

"Right. The plan is to go through that door over there, and into the next building. In that room is a door that opens onto the street. We will open it, and all of us will go out. We will spread out across the width of the street. When I give the order, we will all fire into the backs of those defending the barricade. By that time, Sergeant Blutworth and his lot will be upon them from the front. We will go to work on them with our swords, or musket butts, or whatever comes to hand."

"One hundred," announced Giovanni.

"Good, time to move." Howell led them through the door in the shared wall, and into the ground floor of the next building. When he entered, though, Howell was taken aback. The room, which had been empty only moments before, now held a man with a bloody hand sitting on a bench, and his mate standing over him.

Both of their muskets were propped against the wall. The men looked up quizzically.

Howell did not pause. "Kill them!" He made for the one who was standing. The ceiling was too low for him to get his musket high enough to put the butt into the man's face, but he was able to drive it into the fellow's stomach. The Pilsener tumbled onto the floor. Another of Howell's men flashed by, and pounced upon him. He held a knife, and he plunged it into his enemy repeatedly. The seated, wounded man went down under a storm of blows as Howell's squad filed in behind him and each made his contribution.

"Two hundred," said Giovanni.

The door to the street stood ajar, so Howell stepped over to it, and took a quick glance outside. "There are a couple of dozen of them out there." He kept his voice in a whisper as he stepped back in.

"What?" Howell's men gasped. They were grossly outnumbered. "Dozens?"

Howell closed the door. "It is not as bad as all that." His tone was reassuring. "The reason that the Sergeant wants us to wait our attack is so he can get together the men he needs to come at them from his side." Or at least Howell hoped this was the plan. When he had heard Blutworth telling his men to conserve their shots, he reckoned that the Sergeant was doing so in order to have the means of making an assault. It certainly made sense to coordinate such moves. Howell would know soon enough.

"Three hundred," Giovanni said loudly.

Howell took one of the dead men's muskets, and handed it to the man who would follow him out the door. "Maybe two shots, eh?" He took the other for himself. He filled the captured musket's pan with some of his priming powder, and cocked his match, but did nothing more by way of preparation. He did not want to chance double-loading the weapon. "Giovanni, you are the file closer."

Howell's amico nodded acknowledgment. "Four hundred."

The others moved themselves into a column between Howell and Giovanni.

Howell stood at the door. "Single file. I will go first. I want everyone to make good time out the door, but do it quietly. The bigger the surprise, the better our chance." His warrior lust was rising fit to choke him now. Half of him wanted to run slavering out the door and wring the life out of the first thing he found, and half of him wanted to see his command move efficiently into position and execute a stunningly effective surprise attack from the rear.

"Four hundred and fifty."

The best Howell could hope for was that Blutworth would begin the festivities, and fully engage the attention of everyone on the barricade. Then their surprise would be complete, and Howell and his mates would have the best chance for survival.

"Five hundred."

Howell hesitated. He waited to hear the sounds of Blutworth's attack before he broke cover, but the monotonous rattling of the musketry did not change. He took a deep breath. Someone had to go first. He pulled the door fully open, and stepped out into the street. He and his comrades were only a dozen paces behind the defenders.

He glanced both ways as he walked nonchalantly to the opposite side of the street. It was eerily ordinary, cobblestones under his feet, the street curving away, lined with ordinary brick and half-timbered buildings forming the walls of a major thoroughfare. Shop signs jutted out offering the usual goods and services. Directly ahead of him the tall stoop of a substantial house led up to a doorway at head level. A woman sat on the front step of another house with a bucket and a ladle before her. Doubtless she was there to provide cool water to the city's men as they repelled the invaders. She looked up at Howell and his crew with passing interest.

Lining the improvised barricade that blocked off the street were several complete ranks of defenders, two dozen at least, all dressed much as the men that Howell led. Many of those wearing bandoliers were loading. A few of these put their muskets on the parapet, and loosed a shot in the direction of Mansfeld's Regiment. About half of the Pilseners were holding spears, which were totally unmanageable in this tight space. The fact that they were trying to use these unsuitable weapons spoke to the lack of professionalism amongst the defenders. A rather well-dressed burgher wearing a sword stood behind them, watching them work. One of the men at the barricade noticed the new arrivals, and said something to this 'officer'. He turned casually for a look just as Giovanni emerged.

"Five hundred!" Blutworth boomed from the far side of the barricade. This was followed by a great shout from his, men and the sound of scores of feet thrashing through the loose rubble towards the barricade. Muskets crackled. The officer and all of the other defenders turned to deal with this frontal assault.

*Oow! Oow! Oow!* Howell felt himself slipping from the cool warrior chieftain back into the berserker. He aimed his borrowed musket between the officer's shoulder blades. His match was already in the cock, but when he pulled it into the pan, nothing followed the flash of the priming powder. He was not surprised that the wounded man and his nurse had not carried loaded weapons while tending to the injury. At the end of their little column, pans flared and muskets barked as Howell's men discharged their weapons into the backs of the defenders. The Pilseners' shouts and screams rippled through the tight space filled with mingled powder smoke and fog.

Howell pulled out his match and discarded the useless borrowed musket. He hefted up his own weapon. He blew on his match, and cocked it. The officer whipped around, confusion and terror marking his face. Howell flicked open his pan cover, aimed directly at the center of the man's chest, and pulled his cock into his pan. *Flash! Bang!* The officer disappeared behind a ball of smoke.

Having emptied their muskets, Howell's men now prepared to continue the fight with their swords or the butts of their muskets. Before Howell could do anything however, a shrieking phantom erupted from the haze. The woman that Howell had seen sitting on the step with her water bucket now ran directly into him, wielding her ladle like a battle axe. Her impact knocked Howell's musket from his hands, and it clattered onto the pavement. She was immediately followed by a soldier who seemed just as determined upon flight, pushing Howell to one side, and streaking past. Enemies were barely visible in the smoke ahead, and might come upon him at any instant. Howell could not take the time to retrieve his musket, never mind reload it, but instead pulled his sword. He crouched slightly, and prepared for the fight.

Blutworth's men now topped the wall, and rolled into the Pilseners' position. The enemy spearmen abandoned their useless weapons, and fled as one. Two men carrying muskets joined in the escape, and rushed at Howell. One of them tried to put his musket's butt into Howell's face, but Howell deflected the blow with his one forearm, and jabbed in the sword with his other. It was all instinctive. Howell put no more thought into it than swinging an axe while chopping wood. His assailant went limp, and groaned loudly. Before Howell had a chance to consider what he ought to do next, the second Pilsener swung his musket violently into Howell's arm, and threw him hard onto the steps of the tall stoop behind him. Howell rapped his head smartly against the iron railing, and dropped his sword. Fortunately, that was as bad as it got, as the man immediately resumed his escape.

The fight swirled about Howell as he sprawled upon the steps, stunned by the blows he had taken. It was desperate, hand-to-hand work. Giovanni swung his musket like a club, driving a man to the ground. A Pilsener threw a chunk of brick, but Howell lost track of where it went. Howell watched the fighters with groggy detachment. Discarded spears littered the space. A musket fired. Blutworth's men kept topping the barrier, and finally the defenders could take no more. The few that remained ran for their lives, or threw up their hands to beg for mercy.

*Hurrah!* The Mansfelders congratulated themselves. *Hurray!* "Victory!" Howell saw two that were laughing themselves to tears, slapping one another on the back. "Run, you bastards!" Some sat wherever they could, a look of profound exhaustion on their faces. The small group of prisoners regarded their new status with resignation or fear. Wounded men from both sides lay scattered about, some groaning, some crying, some with men helping them, and others bleeding in quiet isolation. The man that Howell had stabbed was motionless on the pavement before him. A knot of Mansfelders rambled by, sharing a bottle, celebrating their success and their survival. The day's work was far from over, there would be other barriers to overcome and more casualties suffered, but for now they had tasted success and survival, and were content to savor it.

One of the Regiment's men suddenly towered over Howell. "You there, what's the password?"

Since the combatants on both sides were dressed in the same sort of clothes, the only way to recognize a friend was by the password. Howell looked up at the man stupidly. In all his exhilaration and lust for the fight he had completely forgotten the magic words for the day. The musketeer raised his weapon, and aimed the butt at Howell's head. Howell was about to be bludgeoned to death by a comrade because he was ignorant of a simple phrase. "Pay day at Nurnberg," he blurted out.

*Eh?* The Mansfelder paused. It was not the password, but he remembered the incident well. Who would say that but a member of the Regiment?

"Crossing the Alps."

Howell's interrogator relaxed a bit at this other shared recollection.

Howell began to struggle to his feet. This, however, was a mistake. Sitting on the steps, he posed no threat to the musketeer, and his incorrect 'passwords' were interesting enough to give the man pause. As Howell stood though, he forced the musketeer to choose between striking him while he was still vulnerable on the steps, or facing this big man after he had gotten to his feet and was a serious menace. The Mansfelder drew back his musket in preparation for the deathblow.

"Gott mit uns!" Giovanni hurried to Howell's side. He put his body between the musketeer and Howell. "God with us. Mio Dio, he's on our side." The Mansfelder lowered his weapon reluctantly, and then wandered off to resume celebrating with his mates. Giovanni helped his battered friend regain a comfortable seat on the steps. "Are you well?"

Howell thought so, but he had plenty of lumps and pains to consider so he did not answer immediately. The cold fog held down the powder smoke, filling the street with murk and uncertainty. Howell had killed before, but now he had commanded other men to kill. The concept of leading some of his comrades had just appeared in his mind, fully formed. He had discovered yet another aspect of his battle persona.

"Porca miseria, Howell!" Giovanni hovered over his friend. "I thought you were bold standing up to Vlok when we were in that initiation gauntlet, but Gesu Cristo! I do not know where you get this, I do not know what you call it, this crazy-for-fighting thing from but it is, well…"

Sergeant Blutworth stomped up to Howell, and gave him a hearty back slap on his injured shoulder. "That was brilliant, Davidson! Dead brilliant! You are some kind of wunderkinder." He made sure that the 'wonder child' was looking him straight in the eye as he took Howell's hand in both of his. "Well done, Davidson, well done. I wish I had a dozen more like you. Two!" He smiled for a moment, gave Howell's hand a pump, and then moved off to continue his duties. "Dead brilliant!"

Howell smiled up at Giovanni. He had faced death, and survived. He had killed, not just at a distance with his musket, but with his victim's body pressed against his, with the man's smell in his nose. He had seen through the fog of war, and devised a cunning stratagem. He had gathered men and led his hastily assembled command to make his plan actually work. He had been praised for his capabilities by the best soldier he knew. He had felt himself a blood-crazed beast in a fight and found that he could also be a cold, detached god of war as well. *Oow! Oow! Oow!* He was invulnerable, invincible, and immortal..

# Chapter XIV

## House of Three Sisters

Howell sat on the stoop, recovering himself, and looked about at the hazy position they had just captured from Pilsen's defenders. The fog still drifted about, mixed with powder smoke. One of the men that he had led into the attack on the enemy's rear was lying dead upon the cobbles. Howell hardly knew the man. A few of the defeated were being taken into custody. Most of the Mansfelders were gathering themselves, and moving out to force their way deeper into Pilsen. More of Mansfeld's Regiment continued to appear irregularly on the top of the rude barricade, and jump onto the near side.

"All right." Giovanni stood, and eyed his battered friend as Howell sat upon the stoop. "Let's see how our young hero is fairing."

Howell submitted to his friend's inspection, so he almost missed recognizing the man who now stood upon the barricade. Howell paused to look at him more closely. He appeared to be one of the men who had pushed Howell and his friends into the end of the gauntlet at the initiation ceremony. While Howell watched, the ex-Corporal himself climbed up. The toady pointed into the captured position and Vlok raised his musket to fire. What in Hell was he doing? It almost looked like Vlok was aiming right at Giovanni and himself! *Poof! Crack!* A cloud of smoke burst forth immediately in front of the Vlok.

*Ugh!* Giovanni grunted and toppled onto Howell, flaccid as a half-filled sack of grain. He did not make another sound nor another movement. His hat was gone and his hair was a bloody mess. The two of them lay up on the steps of the house in a shapeless pile.

Howell was disoriented. He had never prepared for a moment like this. He had known that his friend could be hit, of course, but it was an abstract thought, theoretical. Now it was real, very, very real.

Then that word 'friend' struck him like a blow to *his* head. Giovanni was his one friend in the whole of the world. In all this morass of greed and battle and treachery and living like a tramp, Giovanni had stood by him. The thought of being denied the one man that he felt was his brother filled him with fury.

Howell sat on the steps with his friend's bloody head on his chest and black rage in his heart. Vlok would pay dearly for this, but first, Giovanni must be tended to. The first thing needful was shelter. Howell struggled to his feet, clutching Giovanni's collar in one of his hands. He pulled his friend up the steps he had been sitting upon and up to the door of the house. He tried the latch. It worked, but the door wouldn't budge. He plowed full tilt into the unyielding door. He should have expected it to be blocked, but he was not doing much thinking now, he was just reacting. Bull-like, he rammed his shoulder into the door. It yielded only slightly, filling Howell with towering dudgeon. He reared back, and flung himself at it with an inhuman fury. It cracked open about a foot. Howell grunted and snarled, ramming himself against the door over and over, dragging his friend the while. At last it was just wide enough, and Howell squeezed through. He wouldn't let go of Giovanni, and he seized his collar with both hands. *ahhhhhhhh!* Finally, Giovanni popped into the room.

Howell sat exhausted on the floor with his friend on top of him. The light from the half-open door and the cracks in the shutters that covered the windows was enough that he could see that he was alone in an ordinary, if rather well-appointed, front room. He was relieved that he had not dragged his friend into a nest of enemies, and, as he relaxed, his desire to rest threatened to become every bit as irresistible as his recent urge to battle. But no! He had to tend to Giovanni!

Howell scrambled up, and laid Giovanni carefully on the floor. His face was covered in blood. It was running off his head and ear, and forming a pool on the floor. Giovanni groaned loudly. Howell suddenly realized that he had feared that Giovanni was dead, but he had not admitted it to himself. Here was proof that his fear was unfounded. What was he to do? How did one deal with a wound like, well, like whatever wound it was that Giovanni had?

Then another thought forced its way to the front of his mind. What if the treacherous Vlok was lurking about, bent on finishing the job? Hell, they were in the enemy's country and neither of them had his musket nor his sword. What would Howell's efforts avail them if they could not defend themselves? Howell returned to the door. A single wooden wedge between it and the floor had held it closed against Howell's efforts. He kicked the bottom of the door, and it popped free. He went back out onto the street in search of a weapon.

For a moment he simply played the spectator. Neither Vlok nor his vile chum was anywhere to be seen. Most of Mansfeld's musketeers had moved off to resume the attack, but there were still plenty of injured and fearful in their wake, each occupied with his own business. Howell descended the stairs, and collected his sword and Giovanni's musket from the steps. Howell's musket was on the cobbles of the street, next to the man that he had killed with his blade. The dead man might have been sleeping except that there was a certain limpness about him that hinted at the truth. His sagging features, viewed from this angle, somehow did not even look human. Howell gathered up his musket and sword, and retreated to the stoop. Everyone was so intent on whatever it was they were about that no one gave any thought to what Howell might be doing. He ascended to the door without looking back.

Howell came through the half-opened door, and stumbled into the comparative darkness of the house. He froze in horror. His eyes were still adjusting to the change in the light, but it looked as though someone was hunched over Giovanni and had his hands around his friend's neck. How could the bastards be so quick? He had only been gone a moment, but it had been long enough for the enemy to attack Giovanni! Howell dropped everything but one of the muskets, roared, and started towards the figure that was poised over his friend. He raised the musket to strike.

Women screamed. The figure over Giovanni turned around to face Howell, and she was screaming, too. She scooted on her butt across the floor away from Howell. He stopped. The other two women cowered with the first, terror and pleading in their faces. Their arms were held out before them in what would have been a feeble defense if Howell had really attacked them. The first women said something tentatively in a language that Howell thought must be Czech.

He looked around the room warily, and looked again at the three now silent but still staring women. He lowered his musket. "I just want to tend to my friend." He said it quietly and in German. He hoped they understood him. He turned, and propped his musket against the wall. He went to his knees next to the wounded Giovanni, who let out another encouraging groan. What was he to do with this bloody mess?

The first woman returned in a wary slide towards the two friends. She fixed Howell with an intense stare. This 'mare', as Giovanni would doubtless have styled her, had disheveled dark hair pulled back above a green dress, and she kept her eyes fixed upon Howell. As soon as she saw that Howell was more interested in his friend than in her, she immediately scooted the rest of the way to Giovanni and re-applied her rag, trying to stop the flow of blood.

Beyond the nurse was another woman. She was wearing a tight, grey doublet with wide sleeves over a dress of the same color, if somewhat coarser, material. "My sisters. We help him." She spoke in accented German, and pointed at Giovanni. "When you leave. He alone was." Her face, half-lit and half in shadow, revealed a strong jaw and an earnest eye. She appeared to be the oldest of the sisters.

"Thank you." Howell moved back to rest against the wall.

Her hair was not in such disarray as was that of her sister. She was a solid-looking woman, but, incongruously, Howell was struck with how attractive she was, in an efficient, mature sort of way.

She suddenly stood. "Excuse." Without waiting for a response from Howell, she turned and went through a door leading to the back of the house.

This left the third sister, the youngest and prettiest of the lot, sitting attentively by herself. Her jaw was angular, like those of her sisters, but in a subtle way that contrasted nicely with the delicacy of the rest of her features. She had her blonde hair in braids, and these braids were wrapped neatly around her head.

All three of these women had a strangely angelic quality about them. Howell wondered if it was possible to slip through a crack in the walls of Hell and find one's self in Heaven.

The woman in the grey dress returned from what was probably the kitchen with an assortment of rags and a bowl of water.

Giovanni let out a huge groan, and sat up. "Gesu Cristo!" He put his hand into the mess upon his head. He took his hand away, and stared, wide-eyed, at the blood upon his hand. He shifted his gaze onto the nurse with no less amazement.

"Calm yourself, friend. Let them finish."

Giovanni turned with obvious pain to look at Howell. "What in Hell?"

"You were hit out there on the street, and I dragged you in here. I don't think it's bad." The last was said as a fact, but it was really only Howell's fervent hope. Giovanni looked like the very Devil.

"Ja," said the woman in grey who had just fetched the rags, "A wound to head has much blood, but little cut."

"And a damned big headache!" Giovanni sounded incensed.

While the two women began binding Giovanni's wound, Howell's attention returned to the woman farthest back. She was concentrating on her sisters' ministrations to Giovanni, so Howell could observe her at leisure. She was not only the youngest of the three, but she appeared to be younger even than Howell. She rocked forwards so that her face was illuminated by a beam of sunlight which came through a crack in the shutters, and fell upon her blue eyes. She was swathed in a shapeless cloak, which seemed odd indoor attire, but the top peeped open to reveal the white of her neck adorned with a necklace of delicate links. Her gaze strayed onto Howell, and she smiled.

The eldest sister stood, and gave a name to the woman Howell was looking at. "Katrina." Her instructions to her young sister were unintelligible to Howell, but Katrina quickly went into the same back room.

"Facile, tesoro," Giovanni murmured to his nurse. "Easy, darling."

The woman who was cleaning Giovanni beamed at being addressed in this melodious language. She glanced up happily at her sister in the grey dress who was standing in the middle of the room, and then resumed her work.

Howell had been cleaning the cobwebs out of his head, and now took stock of their situation. He had left the door standing open perhaps a third of its width, which was far too inviting to any passersby who might have plunder, or worse, on their minds. He walked over and closed it. He looked about, and saw the well-appointed room contained some stools, a large chest, and a rather big table. This latter was not the usual set of trestles and boards that could be disassembled at the end of the meal, but a permanent bit of furniture. This sign of wealth was flanked by not just one, but two wooden chairs, painted with twisting vines and colorful flowers. Plainly the owner of this house was a man of substance. Howell walked over, and retrieved a bench that he thought would serve the purpose he had in mind. It proved to be just long enough that he could pound it down with his fists between the door and the house's timbering, and make the entrance proof against almost any assault.

The woman in grey watched in silence as Howell used her furniture for construction material. The woman in green threw herself on hands and knees across Giovanni to protect him from being stepped upon.

"Porca miseria, Howell. Didn't I tell you I had a headache?" Giovanni spoke to his friend, but he looked at the woman perched across him at eye level, her breasts almost in his mouth. He stuck out his tongue and lapped the air briefly.

The green-clad woman looked down at her patient, thinking that his unintelligible words might have been directed at her.

"Grazie, signorina." Giovanni smiled up at her.

She smiled as well, and turned her head away in embarrassment.

Howell did not cease his hammering, but he did mutter, "Sorry about your head, mate." Satisfied at last with his handiwork, Howell collected their muskets, and propped them in a corner next to the door to the kitchen. There they would be quickly accessible at need.

It was not only the furnishings of this room that suggested a well-to-do household. The shutters protected a pair of goodly-sized windows with honest-to-God glass in them. He and Giovanni had stumbled into a rather posh neighborhood.

The eldest sister approached him tentatively. "Mr. Soldier…"

"Howell Davidson, ma'am." He gave a little salute to the woman in grey.

"I am Barbara." Her German was accented but intelligible.

"Thank you for your kindness to my friend, Barbara."

"Ja, Herr Davidson." She curtsied ever so slightly, acknowledging his thanks. She was not content with small talk however; there was more on her mind. "What is with battle?"

"I am not entirely sure myself." Howell led Barbara to the windows and peeked through the gaps in the shutters. "Let's see." The battle, at least for this barricade, seemed to be over. The soldiers below were just the wake of the fighting, the malingerers, the wounded, and the dead. Howell saw one of the lads standing bewildered in the street. The boy had been rendered nearly catatonic by what he had experienced, while Howell had gone through the same trials as he, plus he had killed a man at close range, been wounded, seen his best friend wounded, and was still functioning. Howell was not sure that it was something that he was supposed to be proud of, but that's how he felt.

"Yes?" Barbara had a look of consternation on her face. She was having trouble finding the words in German. "Who fight? Why fight?"

"Oh." Howell was brought back to the moment. "You mean who are we, and why did we attack Pilsen?"

"Yes, yes!"

"Well, we are Mansfeld's Regiment…"

"Yes, we hear."

"…and we were hired by Count Fredrick of the Palatinate to protect the Protestant Bohemians from the Catholic Emperor."

"Yes?"

Katrina, the youngest of the sisters, the one with the braids wrapped around her head, now emerged from the back room carrying a kettle in one hand and a stack of bowls in the other. She appeared to be offering soup, but Howell could not catch the scent. She hovered over Giovanni for a moment until the amazon in the green dress took one of the bowls and ladled it full. Then Katrina joined Howell and Barbara, and offered some of the food to Howell. It was cold, which explained the lack of aroma, but welcome for all that.

"The Count," prompted Barbara, "the Emperor?"

Howell shifted his attention from the soup back to Barbara. "Ah yes, well, we are supposed to be helping your Estates, your council or…"

"The Estates, yes. Many great men. They talk and they vote."

"As you say. In any event, we came to help, and you Pilseners fought us."

"Why?"

"Why? Damned if I know. The Estates are the Protestants, and you here are all Catholics, I suppose."

Barbara laughed and translated it for her sisters. "No, Mr. Davidson, we are not Catholics in this house. But the big men in the town hall, they like the Emperor."

"Well, then, maybe it is just a fight to see who has the biggest dick."

Barbara had no idea of what those words might mean, but she persisted with the main task. "What now?"

Howell took a deep breath. "Honestly, I am not sure." He turned to her. The faint light coming through the shutters illuminated her face. Her vigorous efforts in his friend's behalf had dislodged many strands of her hair, and she unconsciously brushed them back while they spoke. Despite being older than Howell and her surprisingly strong jaw, her nose was not very prominent. Howell found the contrast rather appealing. "This is my first battle, too."

He stepped closer to the shutters and looked out through them. The morning's fog was burning off, and one could see patches of blue amongst puffy clouds. The sky was beautiful. It contrasted sharply with the ravaged city below.

"This is the first barricade that we have taken today." Howell was talking over his shoulder to her. "I expect there will be a second and a third."

"But finally no more." Barbara turned to look at him. "Finally your Regiment has all."

"Yes." Howell had been entirely focused on the battle itself. Now that his fighting was over, he finally put thought to the aftermath. It doubtless was the most important thing in the minds of his hosts. "I am told that when the city is captured, the men, well…" He had a hard time telling her what would happen next.

"The soldiers get reward." She finished the thought.

Howell cleared his throat. "Yes. We were told that we could do what we wanted with the city for three days."

She swallowed hard, but did not take her eyes off Howell. "Three days?"

*Agh!* A menacing grunt came from the front door as someone tried to force it open. A man shouted in the street.

Howell shoved his half-emptied bowl into Barbara's stomach. "God's bones! It's started already." Howell returned his attention to the window and could just barely make out a dozen or so men gathering in the street at the foot of the stoop. Some of them had sacks or bundles of plunder thrown over their shoulders or resting at their feet. Most of them still carried their swords. All of them had cups or tankards or bottles or whole pitchers clutched in their hands. Someone that Howell could not see, someone standing just outside the door, was giving them orders.

A group of the more sober ones now carried a heavy piece of timber up the steps and suddenly there was a resounding *thump!* that made the house reverberate.

"Oh, merda!" Giovanni groaned and put his hand to his head. "Stop that shitte," he begged of no one in particular.

Howell looked at the bench he had wedged against the door and it seemed to be holding up quite nicely.

Giovanni's protectress, however, dragged her charge across the floor to be clear of the bench, should it fall.

*thump!* Howell began to relax and enjoy the show. His defenses were going to win the day. *thump!* Howell smiled at how clever he had been.

"God's wounds!" yelled the unseen leader. "Some of you get in that window."

Howell immediately returned to battle mode. The sill of the window he was peering through was a good eight feet above the pavement, and he had been thinking in terms of thwarting a handful of drunken raiders. Such men would be unlikely to bother with such a challenging climb. But this band looked altogether more determined. They began to gather benches and barrels and anything else that would serve their purpose and pile them under Howell's observation post. They would soon have someone wrenching off the shutters, breaking the panes, and climbing in.

His battle lust was upon him. Howell drew his sword, and crouched in front of the vulnerable window. He would skewer the first man who tried to enter. His first blow would undoubtedly fuel the ire of the rest, though. How long could he resist them alone? What if they began tossing in torches through the broken window?

Howell's primeval instincts slid into the background, and something smoother and more logical strode to the fore. Once again, he found himself examining the two aspects to his crisis persona. At times it bared its fangs, but at others it wanted nothing more than to outwit its opponent. Howell could choose to which of these he would surrender his soul on any given occasion. Howell stood up straight, and finished thinking through what was in the offing. If he fought, he would certainly be overwhelmed by their numbers. If he ran, he had have to leave Giovanni and the women to the dubious intentions of this lot. What to do? All eyes were upon him. Was there a third option? He unbelted his sword, and tossed it into a corner.

"Ladies, get in the kitchen, and remain absolutely silent." He raised his finger to his lips. *Shhh* He shepherded them into the back room. Before they got out of sight, however, Howell took the kettle of cold soup from Katrina. "Giovanni, as soon as you hear me start talking to these fellows, I need you to wail and groan like You are about to die."

"There will be no difficulty in that."

A great splintering noise filled the room as one of the marauders jammed a bit of wood under a corner of one of the shutters, and levered it open. Howell pulled the ladle from the kettle and flung it, rattling, into a corner. A broken brick smashed in through the window, taking out two panes and the lead strip between them. Howell stuck his hand into the soup kettle and began slopping it into his mouth. He was deliberately messy, splashing the broth all over his face, and rubbing globs of cooked vegetables into his unshaven stubble. Giovanni and the sisters in the kitchen watched this performance in bewilderment.

Howell had just enough time to empty the kettle onto the floor under the window and toss it into the kitchen before the man who had broken the window stuck in his head through the opening. He beat out another pair of panes, climbed in over the sill, and rolled into the front room. He cursed as he rolled in the glass shards and spilt soup, and carefully got to his feet. Although Howell did not recognize him, the man was undoubtedly from the Regiment, as he wore a sword identical to those that had been issued at Nurnberg.

Howell staggered up to him with his head tilted back to keep a mass of cooked vegetables in his mouth. "You've come! Thank God!"

The man's head snapped around at Howell's awkward approach.

*Ooooahahahak!* choked Giovanni, sitting on the floor behind Howell.

"Are you the doctor?" Howell staggered drunkenly to the man. A second face appeared in the window. Howell thought that he recognized this one as a Corporal from another Company. He was probably the leader who had been shouting orders.

*Ahhh!* screamed Giovanni, "God have mercy!"

Howell launched himself at the intruder with a shambling gait and his arms extended in a quest for help. "The doctor!" he pleaded in a rasping voice. "For the love of God!" The pillager backed himself against the wall right next to the broken window. The rogue Corporal's face peeked through the window just opposite the man's hip.

"No, please no!" shrieked Giovanni. "Porca miseria!"

Howell lurched into the raider, and blew the mushy vegetables into the man's face. *Gahh!* Howell did his damnedest to make it sound like he was retching.

The man twisted away from Howell, trying to regain the window.

*Glak bluch gahk* Howell used his tongue to ladle out the rest of his mouthful in a dripping, lumpy slime onto the man's shoulder and all over the floor. "Do not leave us," begged Howell. "The doctor!"

"Get off me!" The man pushed Howell away roughly. Heedless of precedence, he jammed his foot in the Corporal's face in his haste to escape.

"Saints have mercy!" bellowed Giovanni. "The pain!"

"Nooo," wailed Howell. "Please do not abandon us to this horrible plague. You cannot. The disease will kill us! Please come back."

But of course the man was gone. There was a loud clattering as the pile of boards and barrels disintegrated, and fell onto the pavement. Howell moved slowly to the hole in the window. Men were crowded around Howell's recent guest excitedly demanding the news from him while the man tried frantically to get as far from the house as he could manage.

Howell thrust his head out. "Where's the doctor?" Everyone in the street turned to look up at him. "Don't go! I'm sure we can be cured!" He began a dramatic coughing jag, and finally fell back into the house, still hacking.

He did not have to see to know what happened next. The sound of two score feet running across cobblestones is impossible to mistake. He ventured a peek when the sound had faded. All he could hear was the distant growling of a city being ransacked. Except for the plunderers' collapsed assault platform, the street was empty. A church bell was ringing, but not in a regular rhythm. It was sporadic, like workmen were trying to reposition it. There was singing coming from somewhere not too far away, and shouts and screaming and laughter and the sounds of gunfire from almost every direction.

Howell backed away from the window and looked down at his friend. "You sounded great, Giovanni."

"Meraviglioso!" Giovanni was jubilant. "Brilliant! Wunderbar!"

The sisters tentatively peeked out of their hiding place. Howell felt himself beginning to relax.

"That was an amazing idea." Giovanni sat up and repositioned himself. "I knew I made the right choice when I picked you for my friend."

This reminded Howell that not all their comrades were their friends. "Giovanni." He called out loudly across the room causing everyone present to look at him. Although it was pointless in the present company, he still felt that he needed to make his next statement as a public declaration. "It was not some Pilsener that shot you, Giovanni. It was Vlok."

"What?" Giovanni blinked, absorbing the news. "Vlok?"

"Right after Sergeant Blutworth left and you were bent over me, I was looking around. I just happened to see Vlok's ass-kisser on the barricade pointing at us. Then Vlok raised his musket and fired right at us."

"That rat turd!"

"There's no way to know where he was aiming, Giovanni. It was probably at me."

"That's as may be." Giovanni pointed at the bloody rags tied around his head. "That scassacazzo hit me! Where I come from, you do not forget an asshole who does something like that!"

"So now we both have a vende…"

"Vendetta. Si. We both have a man to kill."

In their vehement exchange, the friends had lapsed into Italian, so the women could only experience the emotion of the conversation. It seemed to have an effect, however. Giovanni's green-clad nurse returned to his side, and seemed anxious to sooth whatever it was that so upset her man.

Barbara watched all this, and then moved to her young sister. She held both of Katrina's hands through the cloak that covered her and whispered quite sharply to her. The latter absorbed this lecture with a look that might reflect resignation or excitement, or perhaps both. Barbara finished speaking, and dropped Katrina's arms. She glanced briefly at Howell, and then stepped back.

Katrina hesitated, seemingly gathering her resolve, and then turned to look directly at the soldier who had just repelled an attack upon their home.

It all became clear to Howell in an instant. The eldest sister, Barbara, was no fool. She could see what was coming. For the next few days they would be at the mercy of any thief or rapist who cared to try his luck at their door. Howell had proved a competent defender, and, well, the Devil you know…

Katrina shed her cloak. They had been wise to conceal her until now, as Howell might have entirely missed the attack if he had had this houri to distract him. Shitte, Giovanni might have bled to death and Howell not notice. Her breasts were deliciously full, bulging out of the top of the white dress that barely contained them. Her hips were enticing and her plain dress subtly revealed the unmistakable curves and shapes that Howell found so irresistible.

Katrina glanced at her sister, and Barbara smiled encouragement. The young blonde sidled up to Howell with a little smile upon her face.

Katrina was beautiful. Katrina was ready to pleasure him, but he knew that she would do it only to keep him in her house as a guard. She was every bit as mercenary as was Howell, but then one had to make do with what one had.

Katrina said nothing, but carefully wrapped her arms around Howell's shoulders.

He wrapped one hand around Katrina's waist, and slid the other up to her breast.

She started licking the soup off his face.

# Chapter XV

## Taken by Surprise

Part of the process of holding the captured city of Pilsen had been making themselves unwelcome lodgers in the homes of the citizens. Howell and each of the other men of Mansfeld's Regiment were quartered in a home in which they ate and slept at the expense of the oppressed homeowner. The houses in which Howell and most of his mates lived were tiny, however. The house in which Giovanni was quartered all Winter was cramped even when it contained only Giovanni and the old couple who were forced to feed and house him. However, it was more comfortable than the other houses occupied by Howell and some others, so it had become their usual gathering place.

"All I am saying is that I think that you need to show a bit more respect for cavalry." Howell reached out for the bottle.

Giovanni handed over the liquor they were sharing. "Cavalry, eh? I know of many things to be concerned about in our line of work." Giovanni smiled. "Musket balls, cannonballs, pikes, fellows with razor-sharp swords. I really do not want to add to the list."

"Still and all, I have been talking to Sergeant Blutworth, and he's told me about how dangerous cavalry can be for foot soldiers like ourselves."

"More dangerous than razor-sharp swords?"

"Most of them carry razor-sharp swords. He says that there are cavalry called Croats. These men wear crazy costumes, but do not let that fool you. They are very quick, and appear out of nowhere to chop up men such as ourselves. They are like so many demons."

"They are not demons…" Giovanni crossed himself. "…and you know most of those stories are crazy stuff, like the shitte I make up."

"The kuirassiers are real enough."

"I suppose that's true." Giovanni reached for the bottle.

Howell took a swig before handing it over. "You are damn right that it is true. Armored from head to toe. Two pistols per man."

"Yes, yes, and the armor's all black." Giovanni had heard it all before.

Now their white-haired host spoke up. "The newest kuirassier regiment has been raised by a Colonel named Wallenstein. Do you know how he got the money to raise his regiment of black Devils?"

"Won a lucky dip?" Giovanni offered.

The old man spoke with moral bitterness. "No, he stole the treasury of the Moravian Estates."

"He what?" Giovanni leaned forward to better hear the old man.

"It's true." The elder was confident of his information. "Last year the Estates in Moravia could not make up their minds about joining with the Bohemians or not, so they raised a force to protect their neutrality."

"And?"

"And they appointed this Wallenstein fellow as one of their generals."

"And he paid them back by stealing the strong box?" Giovanni was enjoying this.

"Well, you may think it a joke." Their host plainly did not. "He took their treasure straight to Vienna."

"I will bet his mother was proud."

Their white-haired reporter would not be put off. “When he presented the coin to the Emperor, his be-damned Majesty was so pleased that he gave some of it back to Wallenstein so he could put together his regiment of armored Devils.”

“I seem to have lost the thread here.” Giovanni looked back and forth between Howell and his host. “Both of you have been telling me how dangerous cavalry is, and I thought we were talking about pistols, swords, that sort of thing. Now I come to find that you are warning me that they are all pickpockets?”

Howell laughed despite himself.

The door of the house burst open, and one of the new recruits came stomping in. He slammed the door closed, and looked about angrily. He shrugged off his cloak, and looked in vain for some place to hang it. “Vlok. Put me in mud. Laugh.” He turned, and it was obvious what Vlok had done. All down one side of the cloak and the leg of the breeches below it was a wide wet stain from his fall into the mud. He could not find anywhere to hang his cloak, so he finally crumpled it into a ball, and threw it in the corner. He sat down next to Giovanni. “I hate that Vlok,” he concluded hopelessly.

This reminder of the object of his hatred settled like a dark cloud upon Giovanni's face. He had spent the whole of the Winter unable to conclude his vendetta, and that was with Vlok indoors where Giovanni could do the deed and be out of sight. Now that the Spring was coming, and everyone was out walking in the open, Giovanni’s task would be even more difficult. He tapped his foot nervously. He took a long pull from the bottle.

Howell sought to divert his friend’s attention, so he asked the man who had just joined them, “Where were you when you met Vlok?”

“In front of big town house on big square.” He reached for the liquor, but Giovanni was reluctant to be parted from it.

“City Hall?”

“Yah, that one. Many persons there. Many talking. They say Counts is coming. They say man Surhorse coming.”

“Counts? Surhorse?” Howell had never been able to make out this man’s native tongue, but it certainly was not German.

Giovanni picked up the interrogation. "You say Counts is coming?"

"Yah, Fredrick woman."

"Ah, the Countess. And Sir Horace?"

"Yah, yah." He brightened. "Surhorse."

God's body. Sir Whore's Ass was coming here. Howell did not know what to think. Could he find a way to murder this man? Amongst all these dignitaries? In a city controlled by the ruthless Mansfeld? Howell did not want to let this opportunity pass, but he had to be realistic. He had no idea what Horace's schedule might be, so he had no idea where he might waylay the man.

"So, the high and mighty are out to see the sights, eh? They must be pretty desperate to visit Pilsen in this weather." Giovanni never got tired of complaining about the cold. "Maybe they'll trample Vlok on their way in."

The notion of the dignitaries traveling for their amusement might satisfy Giovanni, but it was not enough for Howell. Why would the Countess and Sir Horace brave the rain and muddy roads to come here? It could not be just a lark. Howell gazed off into space as he followed a thread through his own mind. He surprised everyone by suddenly musing aloud. "You will recall Fredrick has been paying us to side with the Bohemians in their fight against the Emperor. This makes me wonder." Howell leaned forward, the better to engage Giovanni and get his thoughts off Corporal Vlok. "Is our Count Fredrick just a helpful neighbor, assisting the Bohemian Estates out of the goodness of his heart, expecting nothing in return?"

This discussion of political intrigue brought Giovanni out of his funk, but threatened to plunge their visitor into one. He always pulled a sour face when his two friends talked about things that he just could not understand. He got the bottle, and sucked on it for a moment. Then he brightened. "Go see!"

"Go see what?"

"Go see that Surhorse and Counts and many horses and fancy wagons for you should ride in."

"Coaches?"

"Yah, yah!" He hopped up off the bench, and retrieved his cloak from the corner. "We go. We see." He tugged at Giovanni's arm.

Giovanni give a noncommittal shrug. "Would you like to see a parade?"

Howell had very unpleasant memories of the last time he had watched a knot of self-important worthies trot through a city's streets, but he might just learn something that would help him do for Sir Whore's Ass. Howell smiled. "It is not like there is anything else to do." He got up off the bench, and went to the peg where his cloak hung. A parade was about as exciting as anything that had happened in the boring months that they had been in Pilsen.

"Parade be good," the new man crowed as he finished off their liquor, and led them out into the muddy street.

In this poor quarter of Pilsen, the unpaved way was lined with rundown buildings and closed businesses. As they walked, the friends left behind the sort of neighbors and lodgings that were thought suitable for common soldiers, and moved through nicer and nicer parts of Pilsen. They passed buildings with brick ground floors, and some built entirely of brick. They came to paved streets, and to shops that still traded.

Howell spoke in Italian to Giovanni. "There's one thing that I want to tell you that this fellow doesn't need to know."

"Si?"

"If I get the chance, I will not hesitate to kill Sir Horace."

"Be very, very careful, amico."

As they walked, one thing that never changed was the hollow look of the ever-present refugees. The war had disrupted the countryside, and driven mobs of peasants to swell the ranks of the destitute in Pilsen hoping to find relief. Most found nothing, and either starved or relied upon violence and theft to stay alive.

By and by they found themselves in the street that led from the west gate to the City Hall. There was already a considerable crowd assembled, and they were joined by the sort hangers-on that were drawn to such a gathering. There were beggars and performers and hawkers and harlots.

"Mind your purses." When he was a child, barely able to run, Howell had been the one who distracted the victim while an older boy had done the theft. "This is a perfect place to lose yours." When he had gotten older, he had done the actual robbery himself countless times. Some of the company looked quite respectable, so if the quality believed what they had been told, then Howell felt comfortable in doing so, too. They found a stoop in front of one of the buildings that gave a good view of the street, and began their wait.

"So, amico Howell." Giovanni picked up their earlier thread. "You were saying something about that notorious Good Samaritan, Count Fredrick of the Palatinate?"

"Ah, yes, the kind-hearted Count. Truly, I'd like to believe the stories I hear that he pays our wages, and publishes masses of anti-Catholic broadsides, and engages in all the other pro-Protestant spectacle that surrounds him just because he thinks he's doing God's work."

Giovanni snorted.

Howell switched to Italian. "It is possible, it really is. Perhaps I am being too hard on the man, but let us suppose for just a moment that his reasons are not completely unselfish. Let us suppose that he expects to be re-paid for all his good works and he's sending a spokesman, Sir Horace, through Pilsen to Prague. Once he's there, he'll present the bill to the Bohemian Estates."

"The bill?" Giovanni cocked an eye. "What is it that Fredrick wants in return for all this help he's given?"

"I think he wants to be the King of Bohemia."

Giovanni considered this for a moment. "Last I heard, they already have a King. That would make two, which seems rather a lot."

"Well, yes and no." Howell pulled his cloak tight. "It depends on how you look at it. You have to understand things are done a bit differently in Bohemia. It is not like most places where the King dies, and his eldest son takes over as a matter of course. No, no, in Bohemia the Estates assemble, and they vote on who will be their King. Now the old Emperor and King of Bohemia has just died, and no one's been crowned yet, so, in the strictest sense, they do not have a King."

Giovanni raised a finger to caution Howell about assuming too much. "I have heard this too, but a few years past these Estates of yours promised the Emperor that he would be the next King, no? So now that this old Emperor/King fellow has joined his ancestors, wherever they may be…" and Giovanni crossed himself "…the Emperor gets the job."

"All nice and legal, and I am sure that the Emperor sees it your way, but you seem to forget that last Spring these Bohemians tossed the Emperor's men out of the window of Prague Castle. Then they raised an army to fight the Emperor's troops. And when good Count Fredrick offered to lend them our bloody band of mercenaries, and promised to foot the bill as well, did not they just seize on the chance like a thirsty man on a beer?"

"Si…"

"I think that sends a pretty clear message that the deal with the Emperor is off. They hate this git. They are a pack of various Protestants while this monster Emperor is a flaming Catholic. He's determined to bring the lot of them back to Mother Church or see them burnt at the stake. He's the whole reason for the rebellion."

"Let me see if I understand you, amico. These Bohemians we're fighting for want to rid themselves of the Emperor, and make someone else their King, and you think that Count Fredrick is angling for the job."

"Exactly, my clever friend, and the appearance of his closest adviser in our fair city today is the proof of it."

Now that the reason for the visit had been established, all that remained was for the Countess and Sir Horace to actually arrive, which they did only a few hours later. It was a reasonably impressive cavalcade, with horsemen leading the way, and baggage wagons bringing up the rear. Between these came the three coaches that bore the grand passengers.

As the advanced guard of mounted men clattered by, Howell nudged Giovanni, and pointed at the swords and firearms equipping them. "Cavalry, eh? How would you like to be facing those wolves in the open?" Howell could only imagine what it would be like to have one towering right next to him, sword in hand, and these weren't even armored kuirassiers, but just ordinary light horse.

Their perch was high enough that Giovanni was about eye-level with the horsemen as they rode by, but they were plenty intimidating all the same. "I will give it a miss, thank you very much."

*Ahhh* Howell oozed vindication.

After these riders came the coaches. Leading these was a flashy vehicle, sporting bright colors, and a distinctive, elaborate seal on the door. Several men could be seen within. Any one of them might be Sir Horace, but none of them waved or otherwise made himself noticeable, so speculation was pointless. The second coach was far less impressive, but it too rolled by without generating any enthusiasm from those who lined the route. In stark contrast, the crowd was reacting happily to something going on in the last of these vehicles.

Howell stood tall, looking over the heads and shoulders of those in front of him, and saw that several people were waving from the windows of the third coach. From the delicacy of the limbs and the gracefulness of their movements, Howell could tell that these were women who were greeting the well-wishers. He craned his neck to get a better view. His reward was more than he could ever have hoped for. Sitting on the back bench waving gaily to the spectators was a woman with stunning red hair.

In an instant, her coach was directly opposite where they were standing.

Giovanni saw who was waving from the windows. "Say! Isn't that…"

*Oow! Oow! Oow!*

Rosy looked up in astonishment, and began to scan the crowd. She looked about in confusion.

"The big monument!" Howell yelled. "Tonight! The big monument!"

Rosy's carriage passed. He thought that she looked at him, but he could not be sure that she saw him. He thought she waved acknowledgment in his direction, but perhaps she was just continuing to amuse the crowd.

*Oow! Oow! Oow!*

The Moon was just beginning to rise, and relieve the darkness in Pilsen's main square. Howell was sitting on a great assembly of cut stones that immortalized something that had been of some importance to someone at some time in the past. He had been sitting there since Sundown. In the slowly increasing moonlight, Howell could make out some of the details around Pilsen's City Hall. The carriages and wagons that had arrived today were parked beside it in the square. Howell could imagine what the scene must have been like when Sir Horace and his party were greeted by Colonel Mansfeld and his local vassals that he had appointed to give the appearance of civilian rule. By the time Howell had gotten through the crowd and arrived, however, all that was left to see were the servants unloading the baggage, and the animals being taken off to the stables.

That had been hours ago. Now it was dead quiet. Other than a man with a lantern guarding the vehicles, Howell could see no one stirring. Sir Whore's Ass was almost certainly right before him in City Hall. Howell tried to think of a way that he could get in, find him, and kill that son of a bitch.

The guard with his lantern found a comfortable seat. A gentle rumbling of a city at night lay overall. The Moon rose.

Then two figures emerged from behind the wagons. It appeared that this man and woman had come out of the back of City Hall, and they started deliberately across the square towards a large street that Howell knew led to Mansfeld's headquarters. Then the woman pointed to where Howell sat invisibly in the shadows of the monument. They changed direction, and started directly towards him. Howell immediately recognized Rosy.

"There was some great battle here," Rosy prattled on. "I just know it. Men died everywhere." She led the man closer and closer to Howell. "Stabbing. Slashing. Oh, the blood spilt here."

Her escort was Horace's man who had fought at Howell's side during the pirate attack. He was one of the men Sir Horace had used to search the wagons at Genoa when Sir Whore's Ass had shot at him. He had helped Whore's Ass betray him, and turn him over to Genoa's civil guard. Now, though, this heartless thug could hardly have been more bored by his current duties. "Yes, yes. I am sure You are right. Dead men everywhere. Now can we..."

Howell struck. He put his hand on the man's face, and shoved him hard. As the man fell, Howell stepped forwards, and drove his knife deeply into the man's stomach. His head bounced upon the cobbles, dazing him before he could let out a shout. Howell crouched beside him, pumping his arm as quickly as he could, delivering thrust after thrust. The man lay on the pavement without making a sound. He twitched once, and then lay absolutely still. Howell gazed at him coldly. He wiped his knife on the man's clothes.

"I was right." Rosy looked down at the corpse. "Men have died here."

Howell cut off the man's purse. "It will make people sure that he was attacked by some thief from the alleys."

"Good." Rosy took Howell by the arm, and hurried towards the street that left the square. "I knew you must be here. Everyone knew that the Regiment was here, but I had no idea how I could manage to contact you." She glanced at him. "It was so clever of you to find a way for us to meet."

Howell was glad that his unplanned outburst had worked so well. He knew why he had wanted them to meet, but she seemed to exhibit an unexpected eagerness. "Rosy, where are we going in such haste?"

She looked up at him, smiling at the recollection. "You know that Elizabeth and I are more than mistress and servant?"

"Yes." Howell thought it was significant that Rosy consistently used the Countess's given name. She must be on quite familiar terms with the lady.

"Well, we heard your distinctive greeting today, and she thought it sounded like the same dog she had heard at Nurnberg." She glanced at him with an unexpectedly serious look. "When I explained that it was you, she was amazed. Then I had to tell her of our adventure with the tobacco at the payday fair." She glanced over her shoulder again.

"And what did she think of that?"

Rosy continued in her somber tone. "She thought it was scandalous…"

"Scandalous, eh?"

Now Rosy grinned. "…and that she absolutely had to have an adventure of her own."

Howell nodded.

"She knew that her adventure had to be rather more tasteful than ours, of course, so she has decided upon a masked ball. She was very taken by the name you have given me, so everyone is to assume a false identity. She is to be called 'Miss Albion'."

Howell continued down the street with Rosy on his arm. He knew nothing of formal balls, masked or otherwise, but he could not imagine that he had be welcome with his rough manners and the patched clothes that he wore. "So how do I fit into this?"

Rosy guided them into the entrance of a dark alley, and stopped. She looked up at Howell with a determined smile. "Sir Horace will be attending."

Howell's suspicions were confirmed.

"He will be coming right along this very street."

Howell looked towards City Hall as though he could see the object of his hatred coming even now. He had waited a year to visit retribution upon this cockroach, and this was his chance. He could not suppress a grim smile.

"It will be taken as the work of street ruffians, dear, just like the man you've just dealt with."

Indeed it would. "It seems that I am not the only one here who can be devious." Howell looked down at Rosy.

She looked back in anticipation of what was coming. "I have my own reasons for wanting to see Sir Horace dead. Fredrick has fallen completely under Sir Horace's spell. He has become Fredrick's most trusted adviser, and the Count is following any directions that Horace gives. I will not let that vile creature play Elizabeth's husband for the fool." Rosy looked down the street. "He will not guide the Rebellion over a cliff." She looked back at Howell. "You will be the agent of justice, dear."

Howell was filled with satisfaction. He would conclude his vendetta. He would do the world a great service by removing this cur from its population. He would please his woman.

The waiting now began. One would think that with the ball scheduled for a given time, that there would be many revelers passing, but in fact very few people went by, and many of these were ordinary citizens.

"Sir Horace will almost certainly be accompanied." Howell turned, and looked down at Rosy.

It appeared that the thought had only just occurred to her.

He was right, though. He would be attacking not only Horace, but anyone else who was with him. "Do you have a kerchief or something I could tie around my face?"

Rosy produced something far too elegant for the occasion, but it would have to do. Howell was now disguised as best he could be. "I will be using my knife, so that it will look like a vagabond made the attack. Then you and I will escape down this alley. If those around him offer no resistance, I will spare them, but if someone attempts to stop me or follow us down the alley, they will die."

Rosy said nothing, but seemed to accept Howell's plan.

Several men and groups passed them before Howell and Rosy saw a couple that looked very familiar. Sir Horace's rapier hung at his side. The woman wore a long cloak. They had with them a boy holding a torch.

"That woman beside him is one of the household who often sleeps with him. His wife is in England, so she satisfies his needs."

Howell was not concerned with the bastard's whore. He was concerned with doing everything he could to ensure the desired result. He would wait until Horace had passed him. He would fall upon him from behind, slit the man's throat, steal his purse, and make his escape all within a few beats of one's heart. Howell had his knife in his hand. Horace came closer. He was nearly upon them.

A woman's laugh came from further up the street.

Rosy slapped her hand down onto Howell's wrist. "That's Elizabeth!" she whispered fiercely in his ear.

Howell paused. Sir Horace was exactly opposite.

Rosy seized the wrist of the hand in which Howell held his knife. She used both her hands and all the strength they possessed.

Sir Horace went by and was in the perfect position for Howell to execute his plan. Howell did not move. The focus of his vendetta moved away.

"Thank you, dear." Rosy relaxed her death grip on Howell's wrist.

Howell stood in the alley. He stood, and he thought. He had just given up something he had craved with all his heart. He had held back from revenge and justice and the physical joy of killing someone who desperately needed killing. He almost wept at his lost chance. Why had he done it?

Rosy took her hands from Howell's wrist. She was barely visible in the shadows.

Here was the cause of his squandered opportunity. He had let Sir Horace go because Rosy had asked him to.

The second couple passed them chattering happily. They followed a boy with a torch, too. It was the Countess Elizabeth and some gentleman arm in arm with her. This fellow had a rapier at his side like Sir Whore's Arse.

"Howell, if you had attacked Sir Horace, then Elizabeth's escort would have come immediately to his aid."

Howell visualized standing in the street, a bloody knife in his hand, and a rapier pointing directly towards him.

"I could not let that happen, dear."

She had done it out of her concern for him. Perhaps even out of her love for him. Still with his knife in his hand, he reached out to touch her, and she plunged into his arms. They embraced fiercely. Howell smelt her hair. In the dark of the alley, they could only experience each other by how they touched one another. He felt her breast press against his. She looked up at him, and he kissed her just as savagely as he held her.

She pulled back from him ever so slightly, so that she could compose herself. After only a moment, she disengaged from Howell completely. She moved deftly to his side, and slipped her hand into the crook of his elbow.

"So what happens now?"

Rosy seemed to be caught off guard by the question. "I am not sure. I'd been thinking of what would follow Sir Horace's death."

"And what had you thought would follow?"

"I thought we would attend the ball. I already have a suite of new clothes for you in the room we will use." She looked up at him with a grim smile. "I feared that what you are wearing now would be stained with Sir Horace's blood."

"Yet he lives."

"Yes." Rosy was disappointed both that Sir Horace still drew breath, and that her plan to extinguish his life had failed.

Howell had her hand within his arm, and he stepped off in the wake of the Countess Elizabeth. "Something may yet present itself."

They came around a bend in the street and there before him were the man and the woman who had passed them only moments before. The woman laughed, and turned towards the man beside her. She caught sight of Howell and Rosy out of the corner of the corner of her eye. "Oh, my dear! There you are." Both she and her escort stopped to greet Rosy. The Countess had a striking face, with a high forehead. She was grinning gleefully. "You were nowhere to be seen when we were ready to depart, darling. I imagined that you must have preceded us, but here you are. Were you delayed?"

"Yes, Miss Albion." Rosy winked at Howell to emphasize the false identity that Elizabeth was using. "There was some confusion about the time."

"About the amount of time it would take for you to get ready, you mean," Howell mumbled.

Rosy secretly jabbed her elbow into Howell's ribs.

The gentleman with Elizabeth kissed the back of her hand before guiding the Countess before Rosy and Howell. Like Rosy, she was dressed in a long cloak with a fur-trimmed hood.

Rosy looked at her friend. "Miss Albion, may I introduce Howell Davidson…" *ah hmm* "…Hercules."

Where had this name come from? Howell did not have time to consider this, however. He doffed his hat, and continued to play his part in the formalities.

The man with Elizabeth extended his hand, his face smiling genuinely. "A pleasure to meet you, Sir." Howell returned his hat to his head, and the two men shook hands.

Howell appreciated that he was being treated in this wise by a gentleman, but his real attention was focused on another gentleman. Somewhere just ahead was Sir Horace, and the end of Howell's quest.

"Miss, ah, Rosy is it?" Elizabeth's escort said glancing at her. He doubtless knew her real name, but was cool enough to resume his part in the charade that the ladies had devised for the evening. "Miss Rosy here has spoken of you in glowing terms Hercules, and you certainly appear to fit the name she has bestowed upon you."

Howell imagined that the Countess thought that his ragged clothes were just a part of the festivities. "One may play many roles at need." The role he wanted now was executioner.

*Hmm* The man considered this for a moment, and then led the Countess before Howell and Rosy. "Mr. Howell Davidson, excuse me, Hercules, may I introduce…" He cleared his throat, calling attention to the fact that he was announcing a fabrication, "…Miss Albion." He turned his attention to Howell.

Howell knew how to carry himself when faced with musket fire or a hand-to-hand knife fight, but he was absolutely at sea in meeting a Countess. He whipped off his hat, and stood uncertainly in the street, feeling that all eyes were upon him. Did one bow to a Countess or kneel or…

Miss Albion presented the back of her right hand to him. Howell instinctively kissed it as had the gentleman, and hoped that he had not overstepped his bounds.

The Countess Elizabeth had a charming note of sincerity that was positively captivating. "I am very pleased to meet you Hercules. Rosy has spoken of you in the most complementary fashion on many occasions."

"Any friend of Rosy's is a friend of mine." Howell immediately regretted his cocky reply. It might be interpreted as demeaning to the Countess, suggesting that Howell had more respect for his girlfriend than for the lady before him.

"Well!" The Countess stiffened.

*Oh* gasped Rosy.

The gentleman's face hardened.

Then the Countess seemed to reconsider, and view Howell's banter as yet another aspect of her scandalous adventure. "Cheeky Devil."

"Miss Albion!" The gentleman thundered in feigned disapproval.

The Countess glanced happily at Rosy and then to Howell. "My apologies, Hercules."

"None are required, ma'am." A Countess was treating him like a gentleman. Howell released the Countess's hand, and took a respectful step back. "I imagine that the company agrees with what you think of me."

"Indeed." The man who was escorting Miss Albion cleared his throat. "Excessively bold comes to mind."

Rosy pressed her side against Howell's. She and Elizabeth exchanged a significant glance.

"Shall we be off then?" The Countess Elizabeth looked expectantly around the group.

"As you wish, Miss Albion." The gentleman offered his arm. The Countess took it, and these two happily resumed their journey.

Howell was rooted to the spot, holding his hat, looking at the couple receding down the street. A Countess! He had kissed the hand of a Countess, and engaged in witty conversation with her. Then reality came rushing back. Fascinating as all that was, however, his main task still lay before him. He seized Rosy's hand, and placed it inside the crook of his elbow. They made off in pursuit, deeper into Pilsen's darkness, deeper into Howell's quest for revenge.

"Despite all that Elizabeth has tried to do, Sir Horace has only increased his influence over the Count. I'd thought to put an end to it this evening, but now that abominable Sir Horace is still free to work his mischief."

"I can yet do something about that."

"Do be careful, dear. A darkened street is an advantageous place to conduct an attack. Other places are far less attractive."

Now they came to a substantial home that Howell recognized at once. This building was serving as Colonel Mansfeld's headquarters. He was displaying his wealth by putting a candle in every window. The sound of music wafted out. Howell could hear people laughing, and the cheery *clink!* of glassware. The Countess Elizabeth and the man beside her proceeded directly to the front door.

"That's not for us." Rosy steered him to the side of the house. "We have a spot of business to attend to before we join the gay company." She directed Howell down an alley towards the rear of the building.

Howell ground his teeth. He had been shunted to the servants' entrance all his life, so he should not have expected anything better this time, but he had taken the hand of a Countess! He had conversed with a gentleman of the court! Somehow, he had imagined that he had turned a corner, that his life had risen to some new, higher level. The realization that he was still treated as one of the 'lower classes' bit painfully. Then another aspect of this situation occurred to him. Walking in the front door would have raised awkward questions. Posing as one of the servants coming from the kitchen would serve his purpose much better.

It was quite busy at the far end of the alley, with goods still arriving for the festivities that were in progress. Rosy and Howell entered amongst the bustling servants, and Rosy quickly found the major domo. She had a word with the man. Howell was exasperated. He was eager to seize some platter, walk up to Sir Horace, and bury his knife in the man's heart.

A girl was called out of the swirling chaos of the kitchen staff. She led them up a back stair and into a small, but well-furnished, guest room.

Rosy flipped back the fur-trimmed hood of her cloak, and pulled out the bow that held it around her. She shrugged her shoulders so that the girl could take it, and hang it on a hook beside the door. Then the serving girl set to work building up the embers that were in the grate into a respectable fire.

"Leave us." There was not a trace of civility in Howell's voice.

The girl stood in confusion. "But…"

"Now."

The girl scuttled out.

"I am going downstairs, and kill that fucking Sir Whore's Ass."

"No!"

"He will not escape me yet again."

"Howell, dear, there are a score or more of armed men on that dance floor. You may be stabbed a dozen times before you even reach the door."

"Not if they do not know that it was I who did the deed," said Howell. "I will carry in a platter and hold my knife under it. I will look just like the other servants coming and going as I walk up to Sir Whore's Arse. I will slip my knife between his ribs to pierce his heart, and then step away. He'll suddenly drop to his knees, all his mates will gather 'round to see what's wrong, and I will simply walk away."

"Howell." She cast about for something to stop him. She snatched up a parcel off the bed and displayed it to him. "I have new clothes for you."

"I do not care how I look." He turned to go.

"They will make it easier for you to slip in undetected."

Howell paused. He looked down at the package that she had sent ahead. The clothes he wore now were terribly worn. In truth, a new suite would serve his purpose better. "Very well."

Rosy frantically ripped open the bundle. On top was a neatly folded red garment of a fabric even brighter than the rusty color of the doublet he wore now. Beneath it he could see hose, breeches, and a shirt. She looked at him, pleading in her eyes.

Without a word, Howell stripped off his doublet, and cast it aside. His shirt was next. The muscles and the scars of his battle wounds stood out on his torso in what feeble light the embers cast.

"And if it goes wrong, what will become of me, Howell?"

He looked at her in confusion. "We will escape together of course."

"Will we? You may fall to their rapiers. You may be caught, and hanged. Elizabeth knows full well that I love you. I brought you to her ball, for God's sake. If you kill her husband's most trusted advisor, in the presence of a dozen witnesses, she will be forced to act."

Howell hated to think of what action the Countess might be forced to take against Rosy.

"The best that I can hope for is that they turn me out into the streets to fend for myself. They might have me imprisoned. They might have me hanged beside you." She paused to let the import of her words sink in. "And if we do manage to escape, where will we go?"

Howell saw that it would play out exactly as she foretold. Could he once again let Sir Horace slip through his fingers? Could he take the chance that every step of his plan would go as he designed it, knowing full well that there was a very real possibility that it all might go terribly wrong, dooming them both? He was torn. He wanted neither. How could he choose?

"Stay with me, Howell." Rosy was dressed in an elegant dress for an elegant occasion. Her starched lace collar wrapped behind her neck and continued around her shoulders in two arcs that met well below her luscious breasts. Her sleeves were separate from her bodice, and they were attached by laces. Normally, these would be tied by a serving girl working from behind her back, but Rosy had the bows in front. She had planned for this moment. "Please stay." She pulled out the bows, and took off her sleeves.

Killing Sir Horace in the way he envisioned would make him a marked man. Certainly anyone who supported the Estates would hunt him. Even if he and Rosy went to the Army that opposed the Bohemians, they might reject a man bold enough, or foolish enough, to kill his superior.

"Howell." She pulled out the lashings that held her skirt to her bodice, and let it crumple onto the floor.

Howell saw the irresistible triangle between her naked legs with its compelling red hair. She had said she loved him. Did he return that love? He had respected people, but he was not sure he had ever loved anyone before. He did not know what love felt like.

"Stay with me, and I will do whatever you want." The last of her laces held her bodice closed, and she pulled these out. She stripped off the bodice to reveal a tight corset.

He had to make a decision. His craving for revenge had him around the throat, squeezing him relentlessly.

Her naked breasts confronted him. Their nipples were hard. The corset around her body was like nothing he had ever imagined a woman might wear. It bewitched him. She dropped to her knees, and seized the belt around his waist. She swiftly opened the buckle, and pulled down his breeches. She put her face against his thigh. "Please dear, let me finish."

Howell raised one leg, and then the other. She stripped off his shoes and hose, and left them lying on the floor.

Howell stepped out of his breeches.

She looked up at the man towering over her.

He stood naked with the most desirable woman he had ever known kneeling before him. He inhaled the power in which he was imbedded. She would do anything that he asked of her.

"Anything you want, dear."

Anything he desired. All he had to do was take it. He crouched down, took her under the arms, and raised her to stand before him. Their eyes met briefly, and then Howell put both his hands on her breasts, and roughly shoved her onto the bed.

# Chapter XVI

## Sir Horace's Secrets

Sir Horace walked along one of Pilsen's major thoroughfares in the fading light. Dangerous men hid in the dark alleys around him. One of his guards had been killed not far from here, and he was no easy mark. Under such circumstances, Horace would normally have had a servant carry a torch for him, and his other thug by his side. That would not do tonight, however. Tonight Horace was going to meet someone whose very existence needed to remain a secret. Horace would do this by himself, even in such a physically threatening land, which was not all that different from the political isolation that he had created for himself with his intrigues.

He stepped into the public room of an inn. The place was positively teaming with a riotous crowd. Two boxers were beating one another to a bloody pulp in the midst of this mob, and were drawing all of the attention. Sir Horace made his way around this boiling rabble, and walked down a corridor beyond. There were doors on either side, and Horace stopped in front of one of them. He knocked.

"Come."

When he opened the door, he found a man that he had met several times before. The man was stony-faced. He remained seated, and did not offer any greeting.

Horace took off his hat, tossed it onto the table next to the other man's hat, and sat in the only other chair in the room.

The two men stared at one another silently.

At last the other man spoke. "His Majesty is not pleased that you have not yet kidnapped the Count, and delivered him into his Majesty's hands, Sir Horace." The Emperor's man sat stiffly in his chair with an equally stiff look upon his face. "Not at all pleased."

Horace had known that playing a double game would inevitably lead to moments like this. It could not be helped. His first option, becoming the power behind Count Fredrick's throne, had enjoyed initial success. Then obstacles had begun to appear. The Countess gave her husband suggestions that drew him towards paths other than the one Horace had lain out. Many of these thoughts were coming from that red-haired slut who was always at her side. Since his control of Fredrick was not assured, he needed this path to the Emperor to remain open. He needed both of these options, yet he was not yet ready to commit to either of them. He needed time. "It is unfortunate that events have not developed as we had anticipated."

"Not 'we', Sir Horace. You." The agent raised his hands, and formed them into a little tent with all the opposite finger tips touching one another. "You seemed very confident that you could overcome any difficulties encountered when you offered your services to His Majesty."

"Yes, I did, but the situation has changed."

"Changed? How so?" The envoy maintained his stony face. "I understand that you have become quite close to the Count."

"As you say." Horace needed to be evasive with the Emperor's man, but not too obvious. "This is why I am more valuable to the Emperor if both the Count and myself remain as we are," Sir Horace said with complete self-assurance. "How the power struggles amongst the Emperor's enemies will develop is very unclear, and I am now in a perfect position to watch as events unfold."

The Emperor's agent considered this.

"It would be premature to move forwards with the initial plan…" said Sir Horace, "…now that I am far more valuable to his Majesty where I am."

# Chapter XVII

## The Battle of Zablat

The Regiment had been marching for hours, but Howell could see no prospect that they would be able to rest any time soon. Like his comrades, he was oppressed by the endless forest around them, tormented by buzzing insects, and engulfed in the dust raised by thousands of men marching along a crude forest track. His feet hurt. His musket weighed upon him. One thing kept up his spirits, though. They were headed to a city that held out for the Emperor, and that meant a battle.

Giovanni wiped his sweaty brow. "Howell…"

A crackling of musketry erupted from somewhere well before them. Their fat Captain sat up in his saddle, and strained to see the head of the column. Another volley snapped out, and then a third. Where before had been nothing but a column of trudging men and beasts, the Regiment was suddenly alive with shouting, and men looking about themselves in confusion.

"Where are they?"

"Get your matches lit!"

"Who are they?"

An animalistic cry echoed down from ahead. Howell could see the twinkling of drawn swords carried by men on horses rushing in on either side of the head of their column. Muskets and pistols popped. Men yelled. All around him the panic was growing.

"God's wounds! Ambushed by cavalry." The Captain struggled to keep his mount under control.

An aide to Colonel von Mansfeld came roaring by, his horse all in a lather. "Retreat! Fall back and rally! Retreat…"

Confusion reigned. Mansfeld's Regiment fled along the rough backwoods road, hurrying amongst the crowded trees, seeking shelter, and headed they knew not where. A squall of shots and shouts could be heard from somewhere to the rear. Then once from the front, needling Howell that a battle was in progress, and that he was not a part of it. The men around him were scared, confused, tired, covered in the dust they raised, and working furiously to avoid being part of a battle.

"Where we going?" squeaked the pimply little musketeer.

"God's foot," Giovanni snapped at the man in frustration. "For the hundredth time, we do not know. Porca miseria, we do not even know where in Creation we are."

"About a day's march north and west of Budweis." Howell spoke without looking at any of them. He was only half listening to his mates. He was listening intently, however, to the running fight, experiencing the frightening and eerily compelling anticipation of battle. Then he added, almost to himself, "…and nowhere near the fighting."

"And you!" Giovanni turned on Howell. "Your battle blue balls thing is really pissing me off. For some of us…" Giovanni looked about at his road-weary mates, "…all of us really, excepting you, missing out on the chance to get shot at is not all that great a disappointment."

Howell merely sucked his teeth, and continued listening to the firing. It was a galling frustration for him. In the battle for Pilsen he had felt something huge, awesome, and almost unearthly. He had become something else, something more than human, or perhaps less, but unquestionably different, different from the men around him, different even from who he normally was. He chaffed to be that man again.

Another spattering of fire could be heard only a few hundred paces behind them, and Howell turned to see what was happening.

"Give it a rest, won't you?"

For Howell, that was impossible.

The Regiment emerged from the cool of the forest, and into the heat of the open fields around a farming community. A loaded wagon sat immobile in the middle of the road that Howell trod, just where the trees gave way to crops. One of its mules stood exhausted, lathered, head down, and foamy-mouthed. As the friends made their way around this obstruction, they found the wagon's other mule lying motionless on the ground, still in its harness. An angry teamster struggled with the straps and buckles, cursing the dead animal's inconsiderate behavior.

Howell entered a sort of natural amphitheater, with the edge of the forest arcing out on either hand from his location to embrace the little village before them. Bisecting this green-walled arena, the shambling column of the Regiment stretched out ahead of him to the tumbledown settlement another quarter- or half-mile further along the rutted track. The squalid hamlet was rapidly filling with soldiers and horses and wagons and the mass of hangers-on who clung to the Regiment. It appeared that a ring of defenders was being assembled around the village.

"Where this?" No one answered the little musketeer's hundred and first question.

At half the distance to the settlement was a huddle of out-buildings along the road and a double line of pikemen spread out across the fields, their enormous spears standing beside them. These sheds and their Company flag marked the center of a defensive line. This was not the main line of defense being formed around yonder village, but a temporary blocking position. This line did not touch the forest on either flank.

"Oh joy." Giovanni's sarcastic side was on full display. "The welcoming committee."

It was more than welcoming to Howell, it was as a prayer answered. He had heard an hour or more of skirmishing and volleying and shouting all the forenoon, and now he was to have his chance in the fight. His Company was deploying for action.

They halted before the line of pikemen and found a knot of mounted officers on the road before the farm buildings. One of them was Colonel von Mansfeld, small, trim, looking confident, and another was their bulging Captain, massive, sweating, and red as freshly butchered meat.

"Sergeant!" The Captain yelled at them as they approached. "We will defend the town of Zablat." He made it sound like a Greek epic. "My musketeers will deploy here!" He waved his arms across the front as though Blutworth needed instruction as to where to place shotte.

"Yes Sir!" Sergeant Blutworth touched his brim of his hat. He moved along the column, muttering encouragement, gauging the condition of his men, conferring with each of his Corporals as he passed them. Blutworth's limp was unmistakable after the morning of hard marching. He came up along the road to the friends. Howell and the rest formed a line beside him.

Giovanni snapped his fingers. "I have it. It is noon. Must be supper time!"

Blutworth looked at Giovanni like he was daft. "No, del Merico, it is form-the-rear-guard-and-save-the-Regiment's-arse time. Right! Davidson! Take the right flank! Two ranks! Now!"

"A man cannot even ask for a little bread." Giovanni smiled.

He spoke in his usual irreverent way, but out of the corner of his eye, Howell could see Giovanni looking intently at Vlok. Howell wanted Vlok dead almost as much as Giovanni did.

"Right! Two ranks. Let's go." Howell stepped out into the field of half-grown wheat, and led them across the front of the loitering pikemen. The butts of their spears were on the ground while their iron points towered over everyone. Some of them wore breast plates and helmets, but most of them, the bare pikes, had no more protection than Howell. "Halt!" he barked.

The men of the squad ambled into position, with Vlok bringing up the rear. It was high noon on a breezeless Summer's day. Howell's little command came to rest some forty paces out from the road in the calf-high stalks. The insects continued darting here and there around their heads.

Two ranks? Howell considered what this might mean. Their normal formation in an open field like this would be four or more ranks deep. For that matter, the spears should be at least four ranks deep, if not double that. Worst of all, there was nothing to stop enemy cavalry from sweeping around the ends of this sketchy line, and attacking them from the rear.

The Regiment must be desperate indeed to attempt to hold this ground with such a weak defense. Howell and his mates were near the right end of the thin trace of the Company, just two lines of musketeers backed by two lines of pikes, all facing to the southeast. If it came to a fight, they would be hard-pressed to stand with such a shallow formation. The trees on either side of the road before them were easily within range of Howell's musket, but as the wall of the forest arced away from the road where lay the abandoned wagon with its one dead mule, the trees got farther and farther away until they were barely within the distance that the musketeers could shoot.

While his comrades stood idly chatting, Howell continued his reconnaissance. He turned and looked to their right-rear. There was a break in the trees back there where another little road led into Zablat. This opening was in the process of being filled by the men of another of Mansfeld's companies. While Howell and his mates faced southeast, these men faced south, so these two little forces gave a semblance of protection in the directions from which the enemy was most likely to approach. They were only road blocks though, only meant to delay the attackers until the defense of the village was complete.

Howell and his mates settled in for a spot of standing about in the pitiless Sun amongst the hungry insects. A pikeman lost control of his enormous spear, and it glanced off Giovanni as it fell to the ground.

"Inferno! You clumsy oafs are more trouble than you are worth."

"Sorry."

Giovanni was not finished with the pikeman. "At Pilsen you all were as useful as so many lambs."

"This isn't a town, Giovanni. This open ground where cavalry have the advantage. These pikes may prove to be very useful."

Giovanni looked back solemnly, recalling the horsemen he had seen as Sir Horace's cavalcade entered Pilsen.

From the woods ahead of them there was now a sudden eruption of musketry. The irregular crackling and snapping of firearms came closer and closer, and presently a disorderly column came stumbling from the trees. These were the last of Mansfeld's men. The pikes and shotte were mixed randomly with wounded men, men without weapons, and mounted officers. This mob hurried along the track as best they could, chargers rattling on the musketeers' bandoliers, the wounded groaning and dragging their feet. Neither Colonel von Mansfeld nor their Captain was any longer to be seen, so Sergeant Blutworth had inherited the command. He created gaps in the Company's lines so these fugitives could pass through their position, and continue on towards the village.

Howell surveyed what lay before them. "That just leaves us." There was nothing formal about the hand-over of rear guard duty, if one could use the term 'rear guard' for the mass that just gone by. There was simply no one between them and the enemy now, so the honor was theirs.

"I have always liked being the center of attention," said Giovanni.

Presently a smattering of enemy foot soldiers began to emerge from the trees. They made no effort to continue out into the open, but rather filtered to the left along the edge of the woods. They were carrying arquebuses, weapons shorter than muskets, but larger than pistols, so these must be dismounted light cavalry. At this distance, these attackers were otherwise indistinguishable from the defenders, felt hats or knit caps over sweat-stained doublets. Howell thought he saw a few dark patches that might be the marks of blood. He tapped his foot nervously. Was any of them as keyed up about the prospect of a fight as he was?

"Get your matches lit." Blutworth seemed to always be near at hand, giving encouragement or instructions. "They won't come for a spell yet, but best to be prepared."

The blistering Sun nudged to the west while the musketeers shuffled about, starting the ends of their matchcords smoldering.

The rival foot soldiers eyed one another warily. The flies buzzing through the dust motes in the hot air barely registered with Howell.

"Matches, you fools." As always, Vlok was as insulting as he could manage.

An unnatural, clanking, metallic, and otherworldly sound now intruded into the scene. It grew from barely able to hear to impossible to ignore. Everyone along the roadblock watched the exit from the forest, anxious to know what this noise signified.

The first cavalryman who rattled out of the forest was covered from head to foot in black iron armor. He was followed by a steady stream of men similarly equipped. Even their horses were black. Some of these cavalry peeled off to one side, and some moved across Howell's front.

"God's blood," muttered some of Howell's men. "God's wounds," said others.

"Gesu Cristo, Howell! Those are the Black Riders!"

Howell watched the fearsome kuirassiers with a terrible fascination. Each wore a black iron helmet upon his head. He was covered by black iron plates that had been beaten to the shape of his body by smiths like those Howell had seen in Nurnberg. The column that moved across the Company's front glided, and clanked through the field at a majestic pace as they covered the margin of the forest. Each of these riders had a heavy, straight sword on his hip, and two holsters upon his saddle, each of these holding a pistol, so they could fire twice before reloading. Row after row, the black-clad horsemen assembled in the oppressive heat until the whole of the regiment confronted the roadblock in which Howell stood.

"Load." Blutworth walked the length of the Company, calmly repeating that single word. "Load." Their front was transformed into a rippling line of activity.

Howell shook some of his priming powder into his pan, closed it, and blew it off. The eleven powder chargers that dangled from the front of his bandolier were the easiest to reach, so Howell would save them until things got hot. For now, Howell swung his bandolier to gain access to the five bottles on his back, and in particular, the third. He had found that he had the most difficulty getting to that one, so he might as well use it while he was at his leisure. He emptied the contents of the charger down the barrel, taking care to keep the ends of his match well clear of the whole process. He glanced up at the menacing Black Riders. He grabbed a ball from his shot bag and popped it into his mouth. He pulled out his ram rod, and spit the ball into the barrel. Finally, he packed down everything in the barrel, and returned his ramrod to its proper place. He was ready to cock his match and fire.

A single horseman swept onto the field before the line of the Black Riders. He wore the black armor, and rode the black horse as did the other kuirassiers, but he wore a hat instead of a helmet, and across his body was a purple sash that ended in a golden tassel. This was plainly someone special.

"Who that?" A small voice came from one side of Howell.

"Damned if I know." Giovanni was on Howell's other side.

The impressive man skidded to a dramatic halt before the motionless cavalry. He swept off his hat in a bold flourish, and called out loudly, "Men of Wallenstein's Regiment!" His horse pranced about while he brought it under control. The man addressing the horsemen had the pointed little beard that seemed to be universal amongst gentlemen. His hairline had abrupt zigzags that left a pronounced widow's peak above his nose. "The time of glory is upon us!" he proclaimed. The black wall returned a dutiful cheer.

"Giovanni." Howell nudged his friend. "This is the thieving prick that we heard about in Pilsen. He stole the treasury to pay for these Black Riders. It is Colonel Wallenstein."

Giovanni was transfixed by what was unfolding. "Si. Si. The very same."

Wallenstein rode back and forth in front of his Regiment, raising the martial ardor of his kuirassiers with his stirring rhetoric, before waving his hat, and riding off to the cheers of his men.

Howell could see that things might get very serious very soon, so he emulated Wallenstein's action. He stepped in front of his men with a big grin on his face. "That fellow with the fancy sash is wrong. They are not going to cover themselves in glory. They are going to be filled up with our musket balls."

His men smiled, and each encouraged the men around them. They were going to need all the comradery they could muster.

Howell stood in front of them, and completed his preparations where all could see that he was a musketeer just as they were. He put three musket balls into his mouth. When he took the last one out of his cheek, he would know that he had fired four times and that a quarter of his powder was gone.

The cavalry officers across the field barked orders, and their dark troopers obeyed. The front rank of iron-clad men drew their pistols, and fiddled with their mechanisms.

Howell took a position in the front rank of his men. The Black Riders were silent, their horses stomping their feet, shaking their heads, or shivering to dispel the flies. Howell and the rest of the musketeers stood before the kuirassiers in nervous anticipation. The Sun beat down with equal intensity upon the fearful and the clueless and the elated and the drunk. Howell smelt urine as someone nearby wet himself. The heat haze shimmered above the young wheat. Sweat trickled down tense necks, soaking into shirts already laden from the morning's exertions. Someone coughed.

A single shout echoed across the field in front of them. It was picked up, and repeated by officer after officer. Suddenly the black line, stark against the sunny trees behind them, began to trot forwards.

"Pikes!" Sergeant Blutworth strode behind their formation. "Front rank! Against the Horse. Brace your Pike!" Behind the two thin ranks of musketeers, the first row of spearmen bent their left knees, planted the butts of their eighteen-foot weapons against the inside of their right foot, and grasped the shaft to brace it against their left thigh. A comforting hedge of spear points now protruded in front of Howell, their tips about the level of a horse's chest.

"Pikes! Second rank! Against the Horse, Level your pike!" These men remained standing, but held their spears across their waists so that their points stuck out in front of the Company as well. Howell had two pike shafts on either side of him now, leaving him with little enough room to work, but this meant that the cavalry's swords wouldn't be able to reach him. He wished the same was true of their pistol balls.

The kuirassiers trotted towards them in a solid black mass about the width of their Company. They might have easily come at Howell and his mates from the sides, or even surrounded them, but these proud horsemen were sure that they could ride straight over such a feeble defense as they saw before them. They were twenty horses wide and many ranks deep. Their hooves pounded the ground, raising great clouds of dust. Their riders sat impassively, each face surrounded by his helmet, each right hand holding an erect pistol. They might have just emerged from the Devil's own forge.

Howell was buzzing with the anticipation of what was coming. Behind him the pikemen could do nothing except hold their weapons as still as possible and endure whatever the cavalry might throw at them. Only the musketeers in front of them could offer any active resistance.

"Cock your matches!" Bellowed Sergeant Blutworth.

"Cock your matches!" Howell echoed, and he pressed one end of his match into the jaws on the end of his musket's cock.

The menacing block of cavalry was within fifty paces when shouts of command rang out from their ranks. All the pistols in the front rank of kuirassiers came down to point at the defenders. At least half of them seemed to be pointed at the tip of Howell's nose. Unbidden, the sinister thrill of battle once more animated him. *Oow! Oow! Oow!*

"Steady!" Blutworth called out loudly. "Musketeers, front rank, on your mark!"

Howell blew his cocked match into a glowing ember. Then he aimed his musket at the cavalryman directly opposite, and pulled back the cover from his pan. The rider's black iron breastplate jostled as he rode. It seemed that his huge, black horse would trample him in another moment.

"Give fire!"

Howell pulled on his trigger, and drew his match back into his pan. *Flash!* The priming charge went off. *Bang!* His weapon fired next to his ear. Howell and his mates sent their heavy balls flying into the ranks of the approaching cavalry, and added their smoke to the swirling dust that already made visibility so poor.

The reply from the enemy was almost simultaneous. An order was called out, and the cavalry's front rank was enveloped in the crackling of pistols and the smoke from their muzzles. A swarm of pistol balls came humming towards the ranks of the Company.

Given the size of the target presented by the packed infantrymen, it was difficult to miss. Men cried out in pain to Howell's left and right. Pikes and muskets fell to the ground, dropped by wounded soldiers. Howell grinned in macabre fascination.

"Second rank! Step up!" Howell called out.

He happened to see Vlok taking his place in the front rank. He might be a son of a bitch, but he knew what should happen now. He had played the role of musketeer all his life.

"Pikes! Steady!" Corporals all across the line were doing the same.

Howell changed places with the man just behind him, and began the laborious procedure of re-loading. He was all keyed up, but he knew that he had to avoid rushing. Under good conditions, fooling with a bandolier took concentration. There was always the danger of the chargers getting tangled, and now that danger was multiplied because of his exhilaration. Howell forced himself to concentrate, to remain calm, to perform his office. Just do the job. Step by step. Think! How many shots did he have left? The heat and the menacing cavalry and the thrill made it hard to concentrate. You blockhead! There are three balls in your mouth, so you have fifteen remaining.

The front rank of the cavalry peeled off, and began to trot towards the rear. This revealed the second rank of horsemen. They were just as black as the first, just as huge, and just as well armed. They, too, trotted to within spitting distance of the infantry, and leveled their pistols.

Howell had felt vulnerable when standing in the front rank. Now, with the butt of an empty musket between his feet, he felt as exposed as a poster glued to a fence. Fumbling with the bottles on his bandolier, he was defenseless against this attack. He was still a part of the fight, though. It was like he had backed off for a moment in a knife fight.

The front rank of musketeers held their fire, to even out the timing of each of the volleys by allowing each rank the same amount of time to reload. The evil-looking horsemen suffered under no such restraint. They ripped another volley into the infantry, and more cries of pain erupted from around Howell.

"I think I have enjoyed about as much of this as I can stand." Giovanni's face belied his humorous words.

The cavalry was performing a maneuver that Howell had been told was called a caracole. It brought forth rank after rank of pistol-armed horsemen. It could go on and on until the cavalry ran out of ammunition, or they blasted a hole through the infantry. The second rank of kuirassiers peeled off and the endless, deadly procession of the caracole went on. The third rank trotted forwards.

Howell shuffled forward in the column, and withdrew his ramrod. He tamped down everything even though he had yet to straighten his bandolier. The ramrod went back into its slot, and Howell shrugged his bandolier back into place.

The front rank of musketeers volleyed. The weapon held by Howell's partner did not discharge with the others, however. It had missed fire, and he would have to find a way to get his musket working again. He looked scared as he changed places with Howell. Now it was Howell's turn to wait while those behind him reloaded. A horse is a much larger target than a man, so most of Howell's mates had shot animals. Horses lay all over the field before him. Some were just dead. Some were wounded. There was a distinctive difference in how the wounded horses sounded, but they screamed in pain just as men did. Some tried to struggle to their feet, but just fell down again.

Because of the way in which they were dressed, the riders who had fallen off these horses did not look human. It looked like suits of armor staggering about amongst the horses, trying to avoid being trampled.

Another rank of horsemen appeared from the smoke and the dust, like ghosts taking on substantial form. Heedlessly, they forced their way through the disabled horses and their wounded comrades. They aimed their pistols.

Howell's rank fired. He could just make out through the dust and smoke a result of their volley. One of the cavalrymen to their front rolled out of his saddle. It was not quick, but leisurely, as though the man had fallen asleep while riding on a long, forced march. Ever so slowly he tilted over until gravity won out, and he sprawled lifelessly into the dirt.

They were having an effect. They were giving these black Devils something to worry about. They would force the enemy to retire. Howell changed places with his partner. The man looked grim as he took his place. That rank held its fire for one volley from the Black Riders.

Howell emptied a charger into his barrel. His match end burned short, and singed him. He returned his ramrod to its place under the barrel of his weapon. Keep the bloody match lit.

The front rank of musketeers loosed their shots. This time Howell's partner fired with all the rest. The kuirassiers trotted off to the side. The next rank of pistol-wielding shades trotted forwards.

Switch places. Jam the match into the jaws of the cock. Blow the match into a bright coal.

Cavalry pistols were lowered and aimed at Howell.

He brought up his musket. Flipped open the pan. *Flash! Bang!* Howell's shot flew into the enemy's ranks just as twenty more pistol balls came the other way. This time there was not just a yelp of pain here and there from the Company. This time someone just wouldn't stop screaming.

The cavalry before him peeled off. Howell shuffled into the second rank, and began to reload. To his left, beyond Giovanni, he could see a commotion in the ranks and that the pikes were at every crazy angle. The continuous shrieks of pain were coming from that direction. Howell could see a wounded man down there, thrashing about in agony, and disrupting the uniformity of their ranks. If those fellows gave way, they were all lost.

"That bad," was all that the little musketeer next to him need say.

As the next rank of kuirassiers came forwards, one of their number shouted a command. The pistoleers heard this, and aimed off to Howell's left, directly into the area of confusion. They needed but one opening and, all those horses would come pouring through the defensive line, and the riders would commence chopping up the infantry with their great swords from behind. It was terrible to see the effect of a full volley of pistol balls on that small group. There were blood-splattered faces, dropped pikes, dropped muskets, men crawling on the ground, and men lying on the ground unmoving.

Sergeant Blutworth hastened to the scene of the crisis, shouting commands. The screamer was dragged behind the line along with several other wounded. The men stumbled about regaining their positions, Blutworth roaring at them. In a matter of moments, that portion of their hedgerow looked no more filthy, exhausted, nor terrified than any other. Still pumping his ram rod, Howell returned to the front rank, again changing places.

The next rank of cavalry emerged from the smoky, dusty murk, and did not see the vulnerable area. They took no particular aim as they discharged yet another score bullets at the infantry.

*Ahh!* Someone just behind Howell cried in pain. Howell glanced behind him, and saw a man with his eyes wide in shock and his mouth gaping. His feet scrabbled, trying to keep himself upright, but they failed, and he collapsed full length upon the ground. His hands clutched his neck. Gouts of bright red blood pumped out between his fingers, and sprayed all over the shoulder and arm of his doublet. The man looked neither left nor right, but lay twitching and staring into space.

Howell stood helplessly watching. He had never seen someone this close with such a wound.

The head of the wounded man was between the feet of the front rank of pikeman. More blood gushed from his neck, sending a crimson spatter onto the pike butts around him. A final gush of his blood covered the pikeman's shoe in the last of his life.

If Howell could have thought of some way to help his comrade, he would have done it, but nothing occurred to him. Art and faith and magic all seemed equally powerless.

"Gesu Cristo." Giovanni crossed himself.

Others just stared.

What should he do? Howell's mind flipped back and forth between a brother in arms dying before his eyes, and death on a black horse bearing down upon him. What was the proper course of action? What would Sergeant Blutworth do? "Present!" Howell bellowed. He turned back to face the enemy. He was the warrior again. "God's nails! On your marks!"

Another rank of cavalry advanced. Howell hurried to get off a shot at this lot. He cocked his match, and leaned over to take his aim. A random eddy in the dust and smoke swirled an opening through which Howell found his target. As he settled his cheek against the wood of his stock, he suddenly realized that the cavalryman he was preparing to shoot was wearing a purple baldric sweeping across his chest, trimmed in gold, and ending in a golden tassel. This was Colonel Wallenstein himself! The rash fool was inspiring his men by taking his turn in the front rank, and Howell would make it hurt. He hunched over his weapon, and put his finger on the trigger. The image of Wallenstein at that instant would remain etched into Howell's memory for the rest of his life.

The cavalry's volley was so well timed that the individual pistol shots merged into a single sound like tearing canvas. Howell felt someone jerk his hat violently so that it made a quarter turn on his head. Howell's hand twitched, pulling his aim off Colonel Wallenstein as his musket discharged. Furiously, Howell cast about for who had distracted him at the crucial moment, but found that no one was near. He reached up to grab the brim of his hat and put it squarely on his head, but what he found was a severed flap hanging to his shoulder. No one had grabbed his hat, rather it had been struck by a pistol ball that missed his ear by a hand's breadth. Howell froze at the realization of how close he had just come to joining so many others in death. Yet he had not been killed, not even wounded. A feeling of perfect invulnerability flooded into him. *Oow! Oow! Oow!*

Despite his elation at that moment, the battle gradually became almost mechanical for him. The horsemen moved like the hourly display on a town clock, and he and his mates going through the motions of loading and firing like the wheels and levers in a grist mill. It went on like this for a small eternity, but finally a rank of kuirassiers peeled off, and no one took their place.

Howell was too surprised to be glad at first. Then it hit him that they had stood their ground. They had repulsed this attack. "Mansfeld's Regiment!" he bellowed. *Yaaaa!*

Men around him took up the cheer, and soon the whole of the Company was congratulating themselves and cursing their enemies.

*Oow! Oow! Oow!*

"Stand Down your Pike!" The Corporals shouted to be heard above the din. "Stand Down!" Weary and grateful, the men behind Howell planted the butts of their great spears in the dirt, pulled them vertical, and stood at their ease. Now it was over, or at least paused. The whole ghastly regiment of kuirassiers had retired outside of musket shot, but they were not leaving.

Giovanni had caught some of the feeling of success that Howell was enjoying. He glanced about, a great smile upon his face while he stood, fiddling with the things that covered his bandolier, and bobbing this way and that. His gaze penetrated the ranks of pikemen behind them and fell upon wounded men, and dead men, all lying half-hidden in the young grain. Giovanni stopped smiling, but he did not stop fidgeting. At length he found the courage to look upon the enemy cavalry. His gaze changed from apprehensive to quizzical before he finally burst into laughter. *Ha ha!* "Look at those damned wankers, fouling themselves in full view!" His levity was a trifle forced, his eyes a bit wilder than normal, but it was a good accompaniment to the general feeling of success. Several of the men saw what Giovanni saw and began to laugh, as well.

As the murk began to dissipate, Howell looked to see what it was that Giovanni was talking about. The kuirassiers' hands were pumping up and down in their laps. It did look like they were masturbating at first glance. "Sorry, Giovanni, that's not what they are doing."

*Awww* Giovanni made no attempt to hide his disappointment.

"They are reloading." The hard, erect shafts at their waists were their pistols, and the up-and-down movement that had caught his friend's eye was the working of their ramrods. They would be back.

This was not the only thing on the field of military significance. To their left, Howell could see enemy foot soldiers amongst the trees. They had used the distraction provided by the cavalry attack to work along the tree line to the Company's flank. If they kept on like this, they would soon be behind the left flank. This was serious. If Blutworth's men were attacked simultaneously by horse from the front and these foote from the rear, they would be destroyed utterly. He craned his neck, and saw that Sergeant Blutworth was limping his way across the rear of the Company considering all the factors that now contributed to the state of the Company. Most of their powder and ball were gone. Everyone was tired. There were wounded to deal with. They were in a terribly exposed position and had to withdraw as soon as may be. There was not time to deal properly with the dead. They would simply have to be left upon the field.

"By your leave, Sergeant!" Howell waved to get his leader's attention.

Blutworth wiped his face with the back of his hand, leaving a smudge of wet dust across his cheek. He reluctantly shifted his attention, and stepped over to where Howell and his mates stood.

Howell pointed at the enemy's infantry lurking in the trees. "Look there, Sergeant. Quite soon those fellows may come springing out of that wood and take us in flank."

"True." Blutworth scratched his unshaven chin. "You have a good eye, Davidson."

Howell could barely suppress a smile. While his mates were congratulating themselves on not having shot themselves in the foot, or were swatting at flies and complaining about the heat, Howell had been a soldier. He had assessed the situation, and drawn the correct conclusion. "A couple dozen of them at least."

Blutworth made a turn, and looked off towards the other company a quarter mile to their right rear. "But I am equally concerned with what's going on over there." He nodded towards the other roadblock behind them.

Howell twisted about to follow the Sergeant's gaze at their brother company that blocked the road south from the village. They were in some turmoil. Howell realized that he had more to learn about this musketeer business, probably much more.

While Howell and the rest of Blutworth's lads stood quietly awaiting their orders, this other company looked like the surface of a bubbling cauldron. A cloud of cavalry was approaching them from the south. These were not the heavily-armored horrors that their Company was facing, but light, fast, and cunning Croats. They did not form straight lines and a powerful mass like the kuirassiers, but an irregular swarm that flowed over the fields, and threatened to engulf the flanks of the infantry attempting to hold them back.

"They won't stand." The Sergeant gave the scene one more full survey. "We will use this time to get back to the village." He strode out into the hazy field in front of their line. "Company! About face!" The infantry turned their backs upon their enemies.

"Shoulder arms!"

All the powder-stained musketeers and sweaty pikemen picked up their weapons, and placed them on their shoulders for traveling.

"Forwards!"

Howell stepped off, following the pikemen before him, and glanced at the dead they were leaving behind. Howell knew that these would not get a proper burial, but he considered what this meant. What lesson should he learn from this? These corpses lay unburied in the dust of this anonymous field outside this trifling village in a forgotten corner of Bohemia. Howell would not die so. He had told Giovanni to make his name something respected, and Howell intended to do even more. He would be someone. He would be remembered.

"Keep your lines straight!"

Their departure was not before time, either. As Blutworth's men left the field and negotiated a small ditch, the enemy that Howell had seen moving along the margin of the wood emerged from the trees, just too late to get between their Company and Zablat. They loosed a torrent of invective and a few ineffective shots.

Worse, the infantry at the other roadblock had become unnerved by the peril of the Croats swarming about them. Some of these foote began a disorderly retreat towards the village. Some fired, some made their way across the fields, and some discarded their weapons and ran in terror up the road.

"Lively now!" Blutworth eyed the field with concern.

Howell saw that as the panicked company fell back, they left behind what protection the trees had given to their flanks. They were now completely unsupported. Unless Blutworth's men were quick, they too would be cut off by the speedy Croats.

Howell could see a party of these bloodthirsty savages coming at them out of the heat haze from the south. They had colorful scarves around their necks, sported huge black mustaches, and wielded curved-bladed sabers. They yipped and shrieked as though they had just cut their way out of Purgatory. They were still well behind, but unless the Company made sanctuary soon, it would go hard.

Ahead of Howell and his mates, the bulk of Mansfeld's Regiment formed a rough line in the fringes of the Zablat. There were musketeers in front and pikes behind, just as their Company had been formed, but there were many more of them, and the arc of their line seemed to entirely surround the village. Safety was only a few hundred paces ahead.

"They are getting closer, amico." Giovanni's words were unusually grave.

Howell glanced over his shoulder at what Giovanni saw. The other company was approaching the village from the south, and the road they followed converged with the one being taken by Blutworth's men, bringing them ever closer to one another. The Croats rode hard to insert themselves into the narrowing space between the two companies of Mansfelders, yapping and hooting and swinging their swords.

Less than a quarter mile beyond the flank of Blutworth's men, the second company had broken up. The Sergeant looked repeatedly at them as he walked with his comrades. There was no organized formation in that infantry. These foote were streaming through the fields and dashing along the road, herded by a rank of Croat swords twinkling in the Sun as they rose and fell, slashing into the slowest of the infantrymen.

By way of contrast, Blutworth stomped along leading a nearly exhausted Company which he had kept in perfect order despite the strains placed upon it. Howell felt enormous pride as he looked down the length of the well-ordered lines of which he was a part. These silent, weary, but orderly ranks were not just parade-ground bull shitte. This was the perseverance that would keep them alive until they reached the safety of the ring of musketeers and pikemen that surrounded Zablat. The mounted banshees were catching up with them, however. Howell saw that Blutworth repeatedly turned, and walked backwards along the road to keep them in view. Time and distance were against them. A pack of a few dozen horsemen jogged towards them across the fields, and would certainly reach them before they could reach safety in the village. Blutworth turned to walk backwards one last time, and made his decision. "Company! Halt!"

It was not instantaneous. All two hundred men of the Company could see safety tantalizingly close. Every one of them knew that yipping, mustachioed death was hurrying up from behind. For each man, this was a life or death choice. The first pace was at normal speed while this thought, and all its possibilities, shot through their minds. The second step was definitely slowed as the urge to obey their orders took hold. Would the man beside him keep going or slow down, too?

"Halt! Halt!" The Corporals yelled the orders as they moved back and forth before their charges, blocking the reluctant or bewildered men. "Keep your ranks, there!" "You too, asshole!" There were a few whose courage failed them, and they broke for the safety of the village. The Corporals had to let them go. Vlok yelled after them. "Fill your pants, you chicken shittes!"

Howell behaved differently than his comrades. He had seen what Blutworth was watching, and had anticipated the order. He yelled "Halt!" along with all the other Corporals, but he stopped dead at Blutworth's first order. He watched the remainder of the Company recede two or three strides before they came to rest. He did not adjust his position, but remained a lone musketeer, his red doublet stark against the drab background provided by his comrades.

The Sergeant stood between the village and his men. He was ten paces from the line of his men. He looked back and forth at them. He caught the gaze of many of them, and these he deliberately looked directly in the eye, showing his determination. "About face!" he bellowed.

"About face!" The Corporals got their men organized. "Face front!"

This Company might be owned by someone else, and officially bear his name, but when danger threatened, it was Blutworth to whom the men looked. It was their experienced Sergeant who held the men's loyalty, and gave them confidence. It was his lads who got into battle formation, swallowing their fear and facing the menace riding down upon them. They were hungry and sweating and weary and fearful. They were low on powder. Their short line was alone in the midst of a great field. Their flanks were wide open so the bloodthirsty Devils approaching could sweep around them in an instant, and attack them from all sides. Howell knew that continuing the march would have been suicidal. Would this bold confrontation have any better outcome?

# Chapter XVIII

## Escape from Zablat

Sergeant Blutworth had cast the dice. He had stopped his men in the face of perhaps two dozen riders trotting towards them with braided hair and long mustaches above strangely colored scarves and sweat-streaked horses. They brandished their bloody sabers, and yelled like the Devil's drinking buddies. Blutworth had wagered that he could keep these cavalry at bay, and the stakes were their lives.

Howell stood alone several paces in front of the Company's ranks, his musket across his shoulder. He, and the rest of the infantry who stood behind him, regarded the approaching cavalry in silent apprehension.

Blutworth stood in the road behind his men, watching as this menace approached them without apparent concern. Looking between the shoulders of his command as the horse drew near, he called out, his clear voice cutting through the uncertainty that surrounded them, "Pikes! All ranks! Against the horse! Level your pike!" A hundred iron spear points swung down and pointed directly at the approaching riders.

The blood pounded in Howell's ears. His enemy was easily within musket shot now, and he wanted nothing more than to aim his weapon and loose his ball, or to come to grips with one of those fiends like he had done in the alleys as a boy, a sharpened stick thrust under his enemy's nose. No, he told a lie. There was more. He wanted more. It was unaccountable, but even as he beheld mounted death before him, he was thinking about how vulnerable their flanks were, and about how few of their chargers still had powder in them, and whether or not the enemy before him was as tired as he felt. He reveled in the balancing act he was managing between the coolly detached warrior and the primeval beast thirsting for blood. *Oow! Oow! Oow!*

*Yaaay!* His comrades joined in his spontaneous cheering. *Boo!* "Be gone, you gaudy bastards!"

"Davidson!" called the Sergeant across the face of the Company. "Give them a shot."

Howell raised his musket above his head with one hand. "Comrades! I shoot for you!" He turned to face his target. The Company began shouting and waving their hats as he blew upon his match. He cocked it, and flipped open his pan cover. He took his aim at the demon who led the pack before him. He was a big, leering bastard with his enormous mustache covering his broad, crazy grin. Howell squeezed his trigger, drawing his match into the pan of his musket. *Flash!* Nothing followed the flash. The main charge did not go off after the powder in the pan ignited. Howell's musket had missed fire. This failure was not just embarrassing, it might be fatal. He stood up straight, and began to hurriedly re-prime his weapon.

Out in the field, the Croats had reined in, and seemed to be reconsidering where their best interests lay. The body of foot soldiers directly south of the village was in total disarray, easy pickings for practiced murderers such as themselves, while ahead lay unbroken infantry, pikes at the ready.

"Good timing!" shouted Giovanni looking at Howell, and pointing at his musket.

Howell looked back in anger and frustration, but saw his friend was grinning. Howell thought for an instant, and then picked up on Giovanni's bravado and remembered his earlier observation. Howell pumped his fist in front of his crotch as though he was masturbating. *I am just fucking off out here* was the message.

In an instant the entire Company was roaring with laughter.

Howell shook some new powder into his pan, blew up his match, and leveled his musket at the leader once more. He flipped open his pan, and pulled the trigger. This time his musket fired. It was impossible to say if he had hit anything, but he heard shouting from the enemy.

"Davidson, get back here!"

Howell withdrew, but high over his head he held up his fist displaying the universal sign of disrespect, *Fuck you!*

Giovanni copied Howell's display. The whole Company saw him defying the enemy, and joined in the show of defiance.

Here were two hundred unsupported infantry facing a swarm of victorious light cavalry, and the madmen were laughing fit to pee their pants. The Croats paused.

"Musketeers!" Blutworth was going to help the Croats decide what to do. "Front rank. Take your aim!" Fifty firearms were hoisted, and aimed at the threatening cavalry.

The Croats must have decided that whatever was going on, they needn't be a part of it. They peeled off, and redirected their attention to the mob of fugitives upon the south road.

The men around Howell and Giovanni cheered, and whooped. They shouted in defiance and scorn as the Croats sped away.

"Company, silence!" Blutworth needed his men in order.

"Yes Sergeant!" "Right away, Sergeant!" "Silence it is!"

"Rest your pike!" Blutworth watched the retreating cavalry, and then came up behind his men. "Company, about face!"

Howell did as he was bid, and joined the body of the Blutworth's men. The rest of the Company followed their orders without hesitation.

"Company! Shoulder Arms!" Blutworth's bet had paid off. "Forwards, march!" The brave front he had staged had won them just enough time to reach the safety of the ring of defenders that surrounded Zablat. "Company, double time!" His limp was quite noticeable now, and he fell behind, but he continued to call out, "Go! Go! Get in there!"

His men covered the last of the open ground to their lines, panting the last of their wind. The line of defenders raised their pikes slightly and shuffled their feet slightly and created slight gaps in their line through which Blutworth's men shot like rabbits through a hedge.

Howell fell in with Blutworth. "That took a lot of guts back there, Sergeant."

His Sergeant gave a little chuckle. "Guts, eh? I damn near filled my pants with what's in my guts."

Howell gave Blutworth a pat on the back. "Still, if you had not stopped us when you did, we'd all be dead now."

"Well, you lads had to do your part." The Sergeant was limping terribly.

"And who taught us? *Eh? Eh?*" Howell insisted. "You took us boys and made us soldiers." Howell patted Blutworth on the back again, and gazed off towards the little village. "You saved our lives." Howell tried to wrap his arm around Blutworth so that he could help him cover the last few paces, but Blutworth would have none of it.

"I will not have the Company see me give in. You saw the wounded helped off the field, or keeping up on their own. I have asked them stand firm when times are tough. Can I do less?"

Beyond the ring of defenders that Blutworth and Howell had now joined, the fate of the other company was being played out. They had done a terrible job of keeping formation in their attempt to withdraw, and the Croats now had the disorderly mob surrounded. Their sabers thrust and slashed, twinkling in the afternoon Sun. There was almost no musketry, and the chaos of yelling from both sides swirled out in a confused, mingled stream. The Croats whooped and slashed, finishing off any wounded they found.

"Company!" Sergeant Blutworth stood filling, and emptying his lungs like those about him. He raised his sword to make himself as visible as possible. "Rally on me!" Blutworth had found the Ensign, too, so he had the Company's flag to draw the men.

Howell and Giovanni hurried to his side. "Present Sergeant!"

"Lovely. Has anyone seen the Captain?"

"No Sergeant."

"No Sargento."

"Right. Find the Captain and ask him for orders for us. Davidson, you go that way…" Blutworth pointed at the line of defenders in one direction. "…and del Merico you go that way. We will be in the village square, wherever that may be." A place this small might not have a central square at all, but it would certainly not have more than one.

Howell immediately set out upon his search, but he was still vibrating from their narrow escape. He needed to re-focus. He had a new assignment. On Howell's right was the defensive ring set up by the other companies of Mansfeld's Regiment. He passed file after file of tired and confused pikemen and musketeers. They had been on the road the whole of the morning, which was entirely normal on a march day, but this time they had been harassed the while, and now it was well into the afternoon and they still had not been turned loose to plunder their necessities from the locals, so they were hungry and with little prospect of dinner.

To Howell's left, the village was choked with wagons and sutlers and women and boys and all the baggage and parasites that attached themselves to a regiment of mercenaries. Howell directed most of his attention in this direction, as his major goal was to locate his Captain. However, his stomach dictated a secondary objective, so he darted in and out of the houses in the village, and did at least a quick examination of each of the wagons he passed. He found nothing to take the edge off his hunger.

Presently, he stopped to set down his musket so that he could get a bit of a rest, as well. The stock of his weapon rested in the middle of a little farm track even ruder than the one by which they had entered town. It led from the middle of the settlement towards Pilsen, more or less. Two hundred paces out, this track penetrated a solid line of trees at least a half-mile wide.

So far, Howell had seen nothing that even hinted at a gathering of officers. Giovanni was just finishing his travels, and waved as the two came into sight of one another. Howell decided to try to find Blutworth and make his report.

He turned his face towards what he hoped was the center of Zablat, and began to follow the new path he had found. He passed a few poor buildings, but then he saw a more substantial structure just ahead. He had seen nothing better since arriving in town, and wondered if this might not be the place the officers would choose as their headquarters. In any event, he would be a fool if he did not try to discover something edible inside.

He peeked in, and there was Colonel von Mansfeld and some of his aides. One of these lounged near a window munching on a leg of roasted chicken. His Captain was present as well, working his jaw and scowling.

Carefully stacked in a corner of the room beyond him were a great many small, square, iron-bound chests. These were just the sort that Howell had seen filled with coins. Beside them was another, much larger chest, open so that Howell could see that it was filled with papers of every size and shape. Spread over a table was what looked like a map of this very region with Colonel von Mansfeld consulting it intently.

Mr. Bell was seated at a table, quill in hand. He removed his spectacles, and greeted Howell in his usual business-like way. "Good day, Mr. Davidson.

"And good day to you, Bell."

The Captain looked up to see who it was that Bell was addressing. *Hmpf* The note of disgust in his voice was unmistakable. He apparently had still not forgotten his muddied clothes at Genoa.

"Thank God I have found you, Sir." Howell touched his hat brim respectfully. "Sergeant Blutworth has sent me to inquire as to your orders, Sir."

"Ah!" Mansfeld tossed a heel of bread to an eager dog. He did not bother to turn around to address the Captain. "Now that your Company has finally turned up, have them stand in readiness to cut a path for us out through these savages that surround us."

*Hmpf* The beefy Captain glanced at his small Colonel, but the latter did not make eye contact. He returned his gaze to Howell. "Go to Sergeant Blutworth and tell him to make ready for an attack at a moment's notice."

"Yes Sir."

"And be quick about it."

"Yes Sir!" Howell saluted, and ran swiftly out the door. He ran swiftly for about five strides, that is. As soon as he passed the wagon that was parked just outside the Colonel's door, he dropped back to an easy pace, and resumed his search for the Sergeant and for food. Seeing that the headquarters staff had plenty to eat did not improve his humor. He really was not sure of his destination. However, the road that came from the woods and entered the village continued deeper into the town, so he walked along that route.

Everywhere he looked were parked the wagons that bore the Regiment's goods, both official and otherwise. Wagons bearing the tents and other personal effects of the well-to-do officers were mixed with sutlers' wagons and transportation for the Regiment's quota of tarts.

"Well, if it isn't Howell be-damned Davidson!"

Howell looked up to see the French wine seller perched on the box of his blue wagon.

"Riff Raff Ruzeque!" replied Howell happily. "How is it that you have not yet been castrated by Croats?"

"It is against their religion to harm a purveyor of the sacred liquids."

"They were fearful of catching some disease, you mean," Howell yelled over his shoulder as he moved on. He would have loved to exchange quips with this peculiar character, but he did have a mission. Two really. Occasionally, the sound of musket fire or shouting drifted in from their defensive cordon. He poked his head in one tiny house after another, but the village had been thoroughly picked over by the lads who had gotten here first, so he found Sergeant Blutworth before he found something he could eat. His mates were gathered in a dusty rectangle that probably held a tiny market once a month. There was a well exactly in the center, and Howell refreshed himself there. Blutworth limped up to him as he finished his drink. Howell reported. "Sergeant, the Captain is just down this road, in the headquarters of Colonel von Mansfeld. He told me that we were supposed to hold ourselves in readiness for an attack at a moment's notice."

"Right. Davidson, you go back to the Captain."

"Me Sergeant?"

"Yes, you. The Captain will need a runner to tell us when to move, and only you know the way, wunderkinder."

Almost the last thing Howell wanted was to sit at the headquarters, and wait upon his tubby Captain.

Musket fire crackled haphazardly about the perimeter. Shouts and yells issued from the cavalry that seemed to be probing everywhere for a weakness in the Regiment's defenses.

"Now, Davidson."

"Yes Sergeant." Howell shouldered his musket, and began his return journey. Monsieur Ruzeque let him pass in peace, since he was in conversation with the moneylender who had vexed so many at the Nurnberg payday. At least this might spare him any immediate dealings with this poxy-faced leach.

Howell reported his posting as runner to his portly Captain, who was not amused. "Wait outside."

"Yes, Sir."

"And stay on your feet. I want you awake when I call."

"Yes, Sir."

Howell left the headquarters, and stood his musket against the wall. He leaned himself against the wall as well, and tried to relax while remaining standing.

The afternoon slipped by. Things quieted down. Muskets and pistols popped and snapped less and less often. The yells of anger or triumph or defiance or pain became increasingly rare. Howell straightened up respectfully each time one of the officers arrived to confer with the Colonel, and then he resumed his lounging. Various snippets of conversation drifted out to Howell along with the smell of various things that could satisfy one's hunger. The Sun sank towards the horizon. Howell's stomach growled.

Suddenly a blast of angry voices erupted from within.

"And I tell you that your suggestion is pointless!"

"All I am saying…"

"I know what you are saying, and you are wrong. I know General Hohenlohe. He's a fussy, dilatory old woman, and he will not be coming to our rescue!"

Howell's Captain spoke again with a note of wonder in his voice. "How could the fool let five *thousand* cavalry slip out of Budweis when he was supposed to have the town under siege?"

"Exactly. So, now we are the ones surrounded. Waiting here is suicide."

Nothing else that Howell could understand came out to him. He assumed that the only reason that he had heard anything at all was because Mansfeld and the other dignitaries inside had entirely forgotten that he was there. He was there, of course, so his dreary vigil resumed.

Another gentleman approached from between the wagons, and Howell brought himself erect just as he had so many times before. He touched the brim of his hat, and prepared to mumble an ordinary greeting just as he had so many times before. The slight man who approached him wore a large hat with blond curls under the brim. Howell was stunned. For just a moment, he stared at the object of his vendetta.

Sir Whore's Ass waved away a horse fly. The treacherous snake glanced listlessly in Howell's direction, and then away. Immediately he looked back, and sudden recognition froze him in his tracks. His eyes were wide and his mouth hung open.

Howell snatched out his crude musketeer's sword.

In the single heartbeat that it took Howell to arm himself, Sir Horace came to his senses, turned, and sprang into the alley next to the headquarters building. He disappeared so quickly that one might have thought that some conjurer had performed the trick.

Howell dashed down the alley after him. All he could think about was that his bloody revenge was about to happen.

Sir Horace stopped, whipped about, and tried to draw his rapier. The space was so narrow, however, that there was too little room to pull out his sword. He looked at Howell charging down upon him, turned, and ran.

Howell's quarry jumped and dodged rakes and tubs and other household equipment stored in the gangway. Howell must do the deed swiftly and quietly. A fight between a common soldier like himself and a gentleman would never be viewed as an affair of honor, only as murder. Just as Howell negotiated the piled goods, Sir Horace ducked into a back door of the headquarters. Howell made a final leap, and landed in front of the door just as it was slammed in his face. His revenge was just on the other side! The blood pounded in Howell's ears. The man who had tried to get him executed for his own crime was just within. The man who had tried to shoot him was behind this door.

Howell's fingers hovered over the latch. He envisioned himself bursting open the door, Sir Horace cowering before him as Howell thrust his sword into him time after time. The piteous cries followed by gargling, choking coughs, and then silence. Howell barking in triumph at justice administered, his bloody sword held aloft. The mad scramble amongst the officers in the headquarters. The demands that he put down his weapon. His arrest. His hanging.

Howell remained standing before the door, his hand only inches from the handle. Inside, he could hear scuffling, and angry words. Colonel von Mansfeld demanded to know the reason for this strange interruption.

Howell bolted down the alley, sheathing his sword as he went. In an instant, he was in the street and leaning next to his musket. He pulled his hat down over his face, and bowed his head as though trying to avoid the lowering rays of the Sun. What could Sir Horace say? 'I was terrified by the sight of your sentry?' No, he could say that there is someone standing outside your door who's wanted for an attempted assassination in Genoa. Perhaps he did, but Mansfeld wouldn't care. He had got bigger problems to occupy his mind.

Howell's problem was that now Sir Whore's Ass knew where he was. Horace no longer thought that Howell had mysteriously disappeared. Howell was no longer an anonymous face within Mansfeld's Regiment. The man that Howell had vowed to kill knew exactly where he was.

Howell's other problem was living with the fact that he was standing just yards away from the man whose life he had sworn to take, and doing nothing about it. It was galling, but Howell was proud of himself for not doing something stupid. If he waited, if he calmed himself, and kept his eyes and his ears open, something might present itself. He might yet have his chance. Perhaps he would receive the boon he so desperately wanted from Giovanni's Saint Justin.

Whatever excuse Sir Horace offered for his bizarre entrance, it did not seem to involve Howell, as the flies continued to buzz, and the Sun continued to set, and no one seemed in the least bit interested in the soldier parked outside the door, inwardly seething. Officers came and went, conversation drifted out the door, and the last glow of the dusk faded away.

Then there was some commotion to the south. Howell stepped into the street, and tried to see what might be happening. It was full dark. The noise continued, and he heard something similar come from several other directions, too. He looked about, and in the gaps between the buildings around him he saw torches held aloft, bobbing and waving around as though someone from the enemy's side was carrying them towards their lines, and not quietly, either. Whoever it was that bore those firebrands was howling loudly.

*Aye yie!* A handful of Croat cavalry, each carrying a bright torch, suddenly appeared before the defenders at the end of Howell's street. Howell's musket was loaded, but his match had long since been extinguished, so he pulled his sword, and sheltered beside the wagon. The musketeers around the circuit of the defenses had probably let their matches go out, as had Howell, and the pikemen would certainly have laid their great spears upon the ground. The defenders had been unarmed when the surprise attack came, and Croats managed to slip through the defenders into the village from every direction. One of the shrieking fiends came roaring up the road directly towards Howell, the flames from his torch streaming behind him.

In an instant, the outlandishly clad horseman was on the other side of the wagon from Howell. He swept his torch back and forth in the thatch that covered the building. A sputter here and there, and suddenly a large part of the roof lit brightly. The Croat's horse reared, and he struggled to bring around the animal's head so he could repeat his attack upon the headquarters building. The horse plunged across the narrow way, passing Howell's hiding place by barely more than a fathom. The Croat did not see him, but also did not seem inclined to tarry. He threw his torch high onto the roof of the HQ, standing in his stirrups and thoroughly exposing his side. Howell leapt up, and plunged his sword into the man's ribs. The rider toppled off his horse with a groan and into the road. The horse whipped about and started out of the town, trampling the Croat, and continued madly out into the night.

Colonel Mansfeld stood in the doorway of his headquarters, clutching his sword, and looking at Howell. It was not in admiration, nor fear. The look on the trim little man's face was more like stupefaction. The torch rolled off the roof, and fell in a burst of sparks in the road between him and Mansfeld. Howell looked up, and handfuls of thatch were burning in a line all down the roof. Across the way, the roof of the first building was burning in an area bigger than a man's body. Howell turned to look at the sky above the rest of the town, and there were columns of flame coming from a dozen other locations.

The Colonel stepped out into the street.

Howell's flabby Captain emerged from the building, sword in hand.

Mansfeld rubbed his hand across his mouth and nose, and looked about him nervously. "I am under the Ban of the Empire. If I am taken, it is the end for me."

"That must not happen," The plump Captain answered his commander dutifully, but said nothing more, and simply stared as though mystified by what was happening.

Mansfeld surveyed the burning village of Zablat gravely. The Croats had started innumerable fires, and there was no prospect of extinguishing this blaze. Mansfeld suddenly grabbed the lard bag before him by the front of his doublet. "Get your Company up here immediately!" He raised his sword, and pointed along the little track that Howell had used to come into the town. Now it would serve as their escape route. "Your men must cut a path for us through the enemy, and to those trees yonder. Once we are there, we can make a break for it."

"Yes, Sir," he responded, but he took no further action.

"I will see to it, Sir!" Howell took off at a run towards Blutworth and his companions. The tumult was indescribable. Camp followers of the Regiment were everywhere. Wagons and their owners choked the streets. Some tried to get their horses back into harness, some tried to load, and some tried to unload their most precious cargo to have with them when they fled afoot. The noise was stupefying.

Howell battled his way through this mob, and finally arrived in the square. None of the buildings here were alight, but all around them the smoke billowed into the night sky, lit weirdly from below. "Sergeant Blutworth!" He searched among his comrades, some of whom pointed the way. "Sergeant Blutworth!"

"What is it, Davidson?" The Sergeant stood in a small circle of men, most of whom were his Corporals. The thought flitted through Howell's mind that he was in charge of the platoon that the detestable Vlok thought ought to be his. Howell explained their orders swiftly.

"Right," Blutworth pointed back along the road that Howell had just used. "Light your matches. Column of march. Make haste."

Giovanni emerged from the swirling mass. "Howell!" He crossed himself. "Thank God you're safe!"

"Well, for the nonce. We are to strike out into the night, and cut a path of escape for the Captain and Colonel Mansfeld."

Blutworth turned from issuing orders, and took Howell by the shoulder. "Lead on. You know the way."

"Yes, Sergeant."

"Company, follow me!" The Sergeant ordered the Ensign to hold up the Company's flag, and show the men the path they were to follow.

Howell began to worm his way through the bulk of the Company towards the head of the column with Giovanni and Blutworth in his wake. He twisted about, and half-shouted over his shoulder to his friend, "I received another blessing from Saint Justin."

"You have satisfied your vendetta?!?"

"Alas, no."

"Lively now!" Blutworth was getting his men into motion even as he followed Howell. "Make haste." All around them the sky had the look of Hades. An open powder barrel caught fire and sent a pillar of sparks and flame into the sky. Muskets discharged on every side. The screaming of the civilians trapped in the town nearly drowned out the yelling of the soldiers mixed with them.

Howell continued to break trail and he was none too gentle about it. "Make way!" He shoved a woman holding a basket ahead of him until he came to a doorway and then roughly propelled her into the building. "Clear off there!" A man was standing in their path between a wagon and a building.

"God's blest mother." He scuttled off until he found an out of the way place for himself.

Howell came to the blue wagon, and found Ruzeque there. He dumped something into the back of his wagon, and leapt in behind, clearing the path barely ahead of Howell and the column of soldiers behind him.

Howell and those he was leading came to the headquarters building. The aides de camp were in the street trying to prepare, and control the horses that the quality would be using for their flight.

Howell spotted Mr. Bell standing uncertainly to one side. Bell was not wearing his spectacles, but his out-thrust jaw was unmistakable. The bookkeeper had been impressed by Howell's ability to read and write and to do sums. It made them a kind of brothers in arms in this largely illiterate Regiment. Bell been lucky enough to travel in one of the baggage wagons until now, but none of those would be making this trip, and he had no horse of his own. If he was to escape, it would be afoot. "Bell. Come with us." Howell beckoned with his hand.

The paymaster eagerly fell in with them.

In a moment they came to the ragged line of defenders who had once again formed a defensive ring around the circumference of the town. Blutworth stepped forwards to clear the road for his advancing Company. "Davidson! You lead the way straight down this road. Take your time, we still have to form behind you, but do not stop for anything."

"Yes, Sergeant."

Giovanni positioned himself beside Howell, and the two stepped off into the smoky darkness with Mr. Bell close behind.

"Arrowhead formation!" Blutworth said it over and over again behind them. As each man emerged from the wreckage of Zablat, the Sergeant shoved him to the left or to the right, trying to maintain roughly equal numbers of pikes and musketeers on both sides of the wedge they were creating. "Shotte in front. Pikes behind. Arrowhead formation. Keep moving!"

Men began to gather behind Howell and Giovanni in the ghastly light of the burning town. Most of these were from the Company that Blutworth led, but some were from the encircling defenses. Some were just people who saw what looked to be a good chance. Howell and his mates were the tip of the spear penetrating into the unknown, and on either side of them was a growing body of soldiers spreading out into a 'V' formation.

Howell was buzzing. He was the leader, he was the warrior, and perhaps he would be Sir Horace's executioner. Mansfeld had said that they were to cut an escape route for the Regiment's leaders, and Horace would certainly take advantage of this. The burning village cast flickering, uncertain light that would hide what he would do. Muskets were firing here and there. Who would know that it had been his ball that tore through Sir Horace's head? Croats with naked blades surrounded them. Who could say that it had been Howell's sword that had sliced open that bastard?

Sir Whore's Ass had seen him, though. He knew that Howell was somewhere nearby. This meant that he would be watching for Howell, making the deed that much more difficult. Whatever the problems, though, this was the best chance Howell was ever likely to get to take his revenge. He had to seize it. Now.

Howell led the slow advance. He was in his element. It seemed that he could pick out every footfall of the men behind him. He felt that he knew exactly where each of them was without looking. There were seven Croats just there off to the left, waiting, watching. Try as he might, Howell could see no glowing matches before him. These fellows generally carried no firearms, so it was likely that the enemy had no infantry. That also meant that Howell could not shoot Sir Horace without calling attention to himself. It would be his sword, then.

The sudden appearance of the Company seemed to have surprised the enemy, and there was no real opposition. The chaos in the village persisted, but out here that noise was fading, and Howell was conscious that the enemy was ominously quiet. The line of trees was fifty paces ahead. There was no sign of anything between Howell and the wood.

"Victory!" A voice called clearly from directly behind Howell. "Victory!" This was immediately followed by the sound of scores of horses' hooves tearing up the road behind him. Apparently, Mansfeld was using this inappropriate word as the signal for his party to make their escape. The men around Howell saw the troop approaching, and moved to clear the road.

The officers were riding pell mell out of burning Zablat, hunched over their mounts, black silhouettes against the bright flames. Howell stepped off the road, and continued his march as did the others, but he now kept one eye on the mounted men. A rider thundered past, and Howell recognized him as one of the young aides-de-camp. The next rider was undoubtedly the bouncing mass of his Captain. Another horse pounded by, this one bearing the small Colonel von Mansfeld.

The battle unfolding around him receded into the background of Howell's mind. He focused on the man he hated, approaching from somewhere behind him. He managed to contain himself just this side of blind rage. Vengeance! Justice! Another horse approached from behind, and somehow he knew exactly who it carried. Howell saw a rapier silhouetted against the flames. He would end his vendetta in the next few beats of his heart. He pulled his sword and, whipped around.

His mind was embedded in a swirling otherworld with the object of his vendetta at the center of it. Sir Horace already had his rapier pointed at Howell, with the light from the burning buildings glinting off its long, thin blade. Howell took a step forwards to gain the momentum he needed to deliver a savage chop into Sir Whore's Ass. He came at Howell at a full gallop with his rapier aimed straight ahead. Howell struck. His blade slid the full length of the Horace's sword, and was deflected harmlessly by the rapier's hand guard. Howell had not delivered a blow to Horace, but he had forced the rapier ever so slightly off its target. It was enough. The needle-like sword missed anything vital, but penetrated Howell's bandolier and doublet and through his flesh. Sir Whore's Ass flashed by, and Howell was whipped around by the blade that impaled his shoulder. Pain exploded from the top of his arm and surged all down his back. He staggered away from the road, dropping his sword and his musket, and slapping his hand onto the wound. His towering fury at letting Sir Horace slip away again almost blocked out the pain in his body.

Blutworth hurried up to Howell, and cursed. "Panicky bastard."

Howell resisted saying that he knew that Sir Horace's attack had not been a mistake. Missing his chance had been the mistake. Letting Sir Whore's Ass escape was the sin. Howell's terrible anger now began to slide into black depression.

Giovanni handed his musket to one of the men passing by. He examined Howell's back. "Let me see."

Howell was not sure what he was supposed to do.

"Move your hand." Giovanni still could not see anything with Howell's bandolier blocking his view, so he shoved it off Howell's shoulder and dumped it in the road. The great strap and its load of powder chargers and musket balls and whatnot fell straight to the ground, and crumpled up around Howell's feet.

"Sorry," Giovanni grunted automatically.

Reluctantly, Howell examined his palm in the light of the burning village. The flickering light of the fires made everything look red, but Howell had no trouble believing that the red substance that covered his fingers and dripped off his palm was his blood.

Sergeant Blutworth pushed them both towards the trees. "For God's sake! Do what you can on the move, but we must not linger." He shouted out to his command in the darkness around him. "There's nothing for it, lads. If we stick together and keep moving, we will make Pilsen yet!"

"Yes, Sergeant." Howell was trying to do his duty, but he was feeling bleary. He could not focus.

Mr. Bell, unnoticed until now, seized Howell under his good arm and wrapped his other arm around Howell's waist. "Here laddie, there's safe harbor ahead. Let's get there, eh?" The two of them stumbled up the road. Giovanni did whatever nursing he might accomplish for his amico while they walked. The trio started off towards the wood, in the direction where they hoped lay safety, and in the direction in which Sir Horace had fled.

# Chapter XIX

## Sir Horace's Deception

"Sir Horace?" Someone knocked gently on the door to his chambers.

"Come."

The maid came in, tears covering her face. "The Queen has delivered her child, Sir."

It was about time. All the Summer and Autumn he had put up with everyone chatting and speculating and generally rubbing their hands in anticipation of the Elizabeth's next child, and now it was finally at an end.

Sir Horace rose, and led his two men towards the rooms that had been set aside for all the mothering that was going on. In truth, this new addition to the King's family did not amount to much. Fredrick already had plenty of heirs, so this one was unimportant. Still, it was the sort of thing that made the ignorant happy, just like Fredrick's meaningless crowning as King of Bohemia this last month.

He heard a baby squall as he came to an intersection of hallways, but turned away from it to continue. Ahead of him, the sound of women crying cut through the air. This wailing was coming from the rooms where Queen Elizabeth had spent her confinement. More crying erupted. A serving girl came scuttling towards him. She had a bowl filled with bloody rags in her hands, and tears running down her cheeks.

Sir Horace stepped in front of her. "What's going on here?"

She looked up at him, but did not speak. She breathed heavily, and worked to hold back additional tears.

Sir Horace's hand shot out, and shoved the girl's shoulder. "I asked you a question, wench."

"Yes, Sir. Yes." She paused and her lip quivered. "The child is dead, Sir. Stillborn, Sir." She could hold back the tears no longer, and burst into sobbing.

Sir Horace looked down at the blubbering drudge. Dead. The brand-new King and Queen would not be able to present their brand-new child to the masses, and create joy and hope in their simple hearts. The Rebellion had gone badly the whole of the Summer and he had been counting on this tiny bit of good news to take the edge off that. Now he had not that card to play.

A baby squawked behind him. He turned around, and looked in the direction in which lay the rooms of that freckle-faced vixen. Perhaps that womb of hers could provide more than just a place for that red-jacketed soldier to relieve his desires.

He turned to his junior aide. "I want all the woman who have attended the Queen rounded up and isolated in the wine cellar. Have them bring the dead infant with them."

"Yes Sir."

"Lock them in and ensure that they say nothing to anyone." Sir Horace now addressed his senior aide. "You'll take the women who have attended Miss Kincade to the wine cellar as well."

"Yes, Sir."

Sir Horace walked down the corridor. Of course he would have to have them all killed. It was unthinkable that every one of those blubbering females could keep such a secret. One would certainly talk.

He opened the door to the redhead's room. Her baby was wrapped in a cloth, and lay nursing on the woman's chest. Her arms were wrapped around the bundle. Several women were in the room, doing this and that.

"Ah!" said Sir Horace. "It would appear that all is well here, my dear."

The new mother returned a look of complete disbelief.

Sir Horace glanced at the other women in the room. "Miss Kincade and I have some things to discuss. Go with my aide if you please."

There was some hesitation. Several of the women looked to Rosy.

"It will be fine," Rosy said to them. "I'd like it though if you came back when Sir Horace is finished."

"Yes ma'am." The servants moved to the door. "Of course, ma'am." Horace's aide herded them out. The door closed.

They were alone. Miss Kincade eyed him doubtfully.

Sir Horace returned a hard stare. "The Queen's child is stillborn." There was no point in trying to put it in soft words, it was a hard fact. "This, however, can be to your advantage."

The woman looked at him in stupefaction.

"It is important that we display the Queen's new-born child to the masses so that they may derive what pleasure they can from this birth. Therefore, we will substitute your baby for that of the Queen's."

Rosy could not believe what she was hearing. "You will take my child? You monster! You…"

"Calm yourself."

"You piece of filth. You crawling, slimy…"

"I said to calm yourself! Your brat will not be taken from you."

The hussy's face changed from furious to unbelieving.

"Someone must nurse the child. Someone must attend to his filthy garments. Elizabeth would certainly give these duties to some menial, why not to you?"

Rosy did calm herself. She thought about the situation. She thought about what must follow such a deception. "Why not just kill me?"

Sir Horace paused. Why not indeed? He would dearly love to proceed along those lines, but it would not serve. "The Queen would object. That sort of turmoil would not further the cause of the Revolution."

"And what about my women, the ones you have so rudely dismissed?"

"They will all be transferred to duties in the King's more distant possessions so that they will cause no trouble with their wild stories."

Miss Kincade knew full well that this was a bald lie. He could see it in her eyes. She must be thinking that there was a tiny possibility that he would fulfill his promise, but it was far more probable that he would bathe his hands in the blood of her attendants. All the women of her suite, the women she had known and worked with and shared their dreams, would be killed. If she did not accept his plan though, they would suffer the same fate, and her son would lose his mother, too.

"Very well," She finally said.

"Good. That's settled, then. The Queen had intended that the boy be named Rupert, so that is what it shall be."

# Chapter XX

## Sir Horace's Chain of Command

"Make haste, man!" Sir Horace's commanding voice called from within his flamboyant carriage. The vehicle rolled into Pilsen through the gate that served the road from Prague, and inside it Sir Horace shivered despite the furs in which he was swaddled. He tapped his foot nervously, impatient for the journey to end, and anxious to finish the business that brought him here from the capital. When this matter was finally settled, he would be free to return his full attention to making arrangements for the kidnapping of that fool Fredrick. It could not be too soon.

The Rebellion had been frightfully mismanaged thus far by the men who had started it, but there was no guarantee that it would continue in that fashion. The fortunes of war were fickle. The Bohemians might suddenly regain some measure of success, and force the Emperor to grant them peace. It was even more likely that the next Summer's campaign would result in the complete collapse of the Revolution. In either case, the Emperor would have no need to hold the Bohemians' King in his hands, and all of Horace's scheming to the kidnap Fredrick would be wasted.

His visit here would at last free his hands to deal with any obstruction to his plans. The next quarter hour should see an end to this bone-wracking ride in this damnable cold rain as well.

The building that the military dictator of Pilsen had chosen for his headquarters had been the home of one of the richest merchants in the city, but as a Catholic, he had been forced to flee when Mansfeld's men broke into the city. It had been the scene of the Countess's laughable masked ball, and now it would be the scene in which he rid himself of that bastard Dickerson or Donaldson or whatever his name was. Sir Horace was confident that the greedy Mansfeld would eagerly take the money that Horace offered to send the boy to his appointment with Lucifer.

The gaudy carriage rolled to a stop in front of the entrance. The coachman clambered stiffly down, extended the step, and opened the carriage door. Sir Horace shed his skins, stepped through the icy rain, and into the building.

The entry was small, with the foot of a stairway occupying a significant portion of what little space there was. Sir Horace stomped his feet loudly on a portion of the wood flooring where the paint had been worn away, and made a great show of trying to warm himself. "As God is my witness, I have never been so cold."

"I am sorry to hear it, Sir." The words were in English, but with a decidedly Scots accent. Mr. Bell stood to one side while greeting him.

This Bell fellow had translated all his conversations in Pilsen. Mansfeld had assured him that Bell had been with him for years, and was entirely trustworthy. Horace hoped that was the case. He handed Bell his hat, cloak, and portfolio.

"This way, Sir, if you please."

The sitting room they entered occupied the rest of the ground floor, and had been where the Countess's revelers had danced. It had a blazing fireplace, several chairs, and a sideboard with a decanter and glasses upon it. As Sir Horace entered, the mercenary-in-chief rose to greet him. "Good day to you, Sir Horace."

"And a good day to you, Colonel."

"Please, Sir Horace, be seated. Bell, give the gentleman some brandy."

The paymaster made busy with the glassware.

The small, thin Mansfeld settled himself into a chair. "I find myself very unclear as to why you have requested this meeting, Sir Horace." Bell translated Mansfeld's opening of the conversation.

"I am here to discuss a financial arrangement."

"The only 'financial arrangements' I want to discuss are the money troubles of this damned Rebellion," Mansfeld said with a sour face. "What other topic do you think that I want to talk about?"

Sir Horace took a sip of his brandy. "Perhaps you would like to discuss a private arrangement."

Mansfeld waited.

"As it happens, Colonel, there is a man in your Regiment who has committed an offense."

Mansfeld laughed aloud. "Every soul in my Regiment has committed one offense or another. I ordered many of them myself."

"This one affected me personally."

"Indeed?"

"When we escaped Zablat, this man blocked my path." Now Sir Horace artfully twisted his attack upon Howell into an attack upon himself. "He had his sword out, and tried to slash me as I rode by." Sir Horace looked deliberately into Mansfeld's eyes. "Without going into any more detail, I would like to negotiate with you a just remedy."

Mansfeld set down his glass. "Do you envision a court martial, or…?"

Sir Horace let Mansfeld's question hang in the air, waited for a pregnant moment, and then finished it for him. "I envision death."

Bell snorted at this melodramatic performance. He followed this with a passable effort to make it appear he had suppressed a sneeze. "Please excuse me, gentlemen." He translated Sir Horace's intense phrase.

Mansfeld considered Sir Horace. "Death? Just like that? No accusations, witnesses, judgment?"

"There would be compensation."

Mansfeld rose, and walked over to the sideboard. He paused a moment, as though he really had some reason to visit the table, and then he turned to Sir Horace. Mansfeld was not very tall, so this made it possible for him look down upon the man who was asking for his help. "Let me be certain that I am perfectly clear on your request, Sir Horace. You propose to give me money in exchange for the death of one of my comrades."

Sir Horace drained his brandy. He looked up at the Colonel. "Yes, exactly."

"How much?"

"Beg pardon?"

"Sir Horace." Mansfeld carefully pronounced his words so that both Sir Horace and Mr. Bell could be in no doubt as to his question. "How much money do you offer me to kill this fellow?"

Horace wanted that bastard sent to Hell, but how much was he willing to pay? There would doubtless be some back and forth on the fee. Best to start low. He looked at Mansfeld. "Five hundred kreuzers."

*Pffft!* Mansfeld straightened up. "I wouldn't kill a dog for so little." He turned back to the sideboard.

Sir Horace hardly needed Mr. Bell to translate this for him. Sir Horace tried again. "Colonel, we are all short of cash at present." This was no more than the truth. The Rebel government of the Estates in Prague had run up enormous debts. Wages had ceased. Bills had gone unpaid.

"Short indeed." Mansfeld spoke with some vehemence. "Do you know how much this Regiment is owed?"

Sir Horace was ignorant of the details of Mansfeld's ledger books, but he knew the financial crisis the Estates were facing. He could make that calculation. "Precisely why I thought cold silver might be welcome here."

Mansfeld looked at Bell as he finished translating. The English dandy with his blonde curls had a point. The Colonel considered for a moment, and then took his next position. "And what about my men?"

"I am sorry?"

"The men who will perform this little task for you, Sir Horace. How much will you pay them?"

Horace had thought that he was negotiating for a complete contract, but now it appeared that this slippery Colonel wanted the five hundred kreuzers just to give his permission, and that he wanted Horace to pay separately for the deed itself. Horace was caught off balance.

"My dear Sir, let us say one thousand kreuzers to cover everything." Mansfeld addressed him like a professor lecturing a promising student who had yet to reach his full potential. Bell made every effort to faithfully reproduce the mocking tone in his English translation.

Sir Horace sat mute, his tongue running over his lips, and his foot tapping. The sum was enough to give him pause, but not so much as to make him refuse outright.

"Now that's settled, what's the fellow's name?" Mansfeld stood expectantly.

"Davis."

The Colonel arched his brows, and looked at him questioningly.

Sir Horace looked away in irritation. First the hucksterism about the money, and now this tutor-like insistence on the full answer. "Howard Davis." Or was it Harold?

Mansfeld addressed Mr. Bell directly. "Amongst the men we've managed to collect for the Regiment, in what Company is this Howard Davis fellow?"

Bell had kept the muster rolls month in and month out for more than two years. He had added and struck out hundreds of names in that period. He had done a thorough census of the men who had survived the destruction of the Regiment at Zablat, and he knew who was meant. His answer was prompt. "Sorry Sir. No one by that name in the Regiment."

Sir Horace sat there with his mouth opening and closing, but no sound coming out.

Mansfeld looked at Sir Horace, and then back at Bell, an expression of resignation marking his features. "Are you sure, man?"

"Aye, positive, Sir. However, we do have a Howell Davidson in Dichburg's Company. He was enrolled in Verona."

"Yes, yes." Sir Horace chirped brightly. "That must be him. Fought the pirates. Offered up in Genoa."

"Offered up?" asked Bell.

"No, no, no," Sir Horace snapped irritably. "Never mind about that. Zablat. He was in the battle at Zablat."

"Doubtless, Sir."

"Very good, Bell." Mansfeld and his guest worked out the details, and then Mansfeld pointed towards the door. "Get Sir Horace's things, Bell."

"Immediately, Sir."

Sir Horace knew that some small talk here would be proper, but as the men had no language in common, it was nearly impossible. Then Horace noticed a sword hanging on a display over the hearth. He stood and walked over to it. He turned to Mansfeld with a look that said, *'May I examine this?'*

Mansfeld swept his hand to offer an inspection. *'Be my guest.'*

It was a fine sword, to be sure, but Horace could read nothing of the inscriptions etched into it. The grip was Morocco leather bound with gold thread, and a golden tassel hung at the butt. A very beautiful weapon…

Suddenly, fear gripped his heart. Where was his portfolio? He kept it with him at all times, so that there would be no opportunity for anyone to see his secret correspondence with the Emperor. All of this discussion about murder and blood money had totally occupied his mind until now. Bell seemed to be taking a surprisingly long time to fetch a cloak and a portfolio. Had the paymaster taken a moment to rifle his papers? "Colonel…"

Mr. Bell walked into the room, his face utterly impassive. He had the cloak draped over one arm and the portfolio in one of his hands. The portfolio looked to Sir Horace exactly as it always did. There was no evidence of any tampering.

Mansfeld walked over to Sir Horace, and took the sword he had been holding. While Horace was getting into his cloak and taking his portfolio, Colonel Mansfeld offered his hand. "It goes without saying that this will remain a closely guarded secret."

Sir!" Captain Dichburg got to his feet as Colonel von Mansfeld unexpectedly appeared in his quarters. Dichburg's house was not so grand as his Colonel's, but it had once been the pride of a successful merchant as well.

"Calm yourself, Dichburg." Mansfeld swept off his cloak, and settled himself into a chair. "We have a chance to make a little money, you and I."

"I am all ears, my old friend." The Captain seated his bulky form next to his commander's thin one. They had been comrades and business partners for many years now, serving together for the Emperor, the Duc of Savoy, and now these miserable, penniless Rebels.

"Specifically, you can make four hundred kreuzers for an afternoon's work."

"I am free this very day."

Mansfeld gave a little laugh. "It may take a bit of planning, so perhaps next week would be better."

"Then next week it is. What's needful?"

"There is a gentleman from Prague, Sir Horace. For the last year or two, he's been haunting the court, first when Fredrick was a Count and now that he's the new King of Bohemia. You know him?"

Dichburg nodded.

"He has a dispute with one of the men in your Company, one Howell Davidson."

"I know this oaf. He knocked me into the mud in Verona."

"Excellent! Then you won't have any objection to killing him."

"Kill him? And four hundred kreuzers, as well? You make me smile."

"There are a few restrictions that must be observed," said Mansfeld. "First, he must just disappear. Like he deserted or wandered off and froze to death or something of that sort. The gentleman doesn't want anything to arouse suspicion that this Davidson fellow was murdered."

Dichburg listened intently. "So no one will wonder who did it, or ask who might have wanted it done."

"Precisely. Second, Sir Horace wants to watch."

"Of course." Dichburg grinned. "There's been so little entertainment of late."

"Correct. So make your arrangements, and then notify Sir Horace. He thought somewhere far out in the countryside would be nice, with you and him meeting in the morning, and then having the performance in the afternoon."

"Consider it done, Sir." Mansfeld rose to leave, and Dichburg stood at the same moment. "And Sir, how much will Sir Horace pay my men?"

"Please, Dichburg, I taught you that trick. Pay what you must from your own pocket, but it is four hundred from first to last. And keep this secret well."

Volk!" Dichburg lumbered into the house where his newly reinstated Corporal stayed. "Vlok! Where are you?" Vlok was quartered in a house that was one of the ordinary, half-timbered variety, but in a nicer part of town as befit a Corporal.

Vlok came bustling out of a side door, a look of submissive eagerness upon his face. "Here, Sir. Here I am."

"So you are." The Captain pulled off his cloak, and tossed it to the man. Dichburg walked into the kitchen from which Vlok had just emerged, and found it served as the sitting room, too. At least it was warm. Dichburg sat his ample frame in the only chair to be seen. "I have a little job for you, Vlok."

"Yes Sir." There was nowhere for him to sit, so he stood, and listened to what his Captain had to say.

"Next Tuesday…" Dichburg began, but he immediately saw that Vlok was bewildered by this way of reckoning the passage of the days. Vlok's sense of time was regulated by his hungers, not by a calendar. Dichburg therefore simplified things, and said, "In three days' time, you will lead a foraging party to the village of Sec." Once again Vlok frowned, so Dichburg explained further. "It is about four hours out on the Budweis road, and maybe a half hour's walk south. It is the one where we got into the fight over the cow a couple months back."

Vlok brightened at this. "Oh, yes, Sir. I know the place well."

"Your foraging party will leave just after dawn."

"Yes, Sir."

"You will take with you two men that I will pick that can be trusted to keep silent, and Davidson."

"Davidson?!?" Vlok repeated the name with such distaste that he totally forgot his manners and did not append 'Sir' to his outburst.

"Yes, Davidson." Dichburg smiled in a way that left Vlok staring in confusion. "Once you get to the village, you will take your party into the big barn."

"Half the thatch was burnt off the last time we saw it."

"That's the one. You will tell your two men and Davidson that you are going on a quick scout. Leave them in the barn, and come to the big house."

"The one with a chimney at each end."

"Exactly. You will find me waiting there. All you have to do is make sure I know you are there. I will tell you to carry on, and then you'll go back to the barn."

Vlok was plainly having difficulty processing such a string of commands. "Begging the Captain's pardon, but…"

"I have not come to the best part, yet, Vlok."

"Best part, Sir?"

Dichburg's evil smile returned. He leaned forward to put his face close to Vlok's. "You will return to Davidson, and take him out into the open space just at the corner of the barn. Then you will kill him."

Now it was Vlok's turn to grin. "Kill Davidson, Sir?"

"Oh yes, Corporal, you will kill that son of a bitch. And just so you and your two friends will have something to encourage you to keep your mouths shut, I am going to give you a hundred and fifty kreuzers to share amongst yourselves."

Vlok wouldn't have been happier if he had just passed through the pearly gates with a girl on each arm.

"And Vlok, do not repeat a damned word of this, understand?"

# Chapter XXI

## Visiting

Howell heard someone rapping on the door of the house in which he was quartered. The sound was not loud, but it was insistent. One set of raps was quickly followed by another. Whoever was outside wanted to be off the street as quickly as possible. Howell pulled open the door.

"Howell." Rosy pushed past him and into the room. "Thank God you're here."

This was completely unexpected. He stepped out into the cold night, and looked both ways. There was no one to be seen. He came back in, and closed the door. "For Christ's sake, Rosy. How do you come to be here?"

"I walked." She pulled out the bow that held her cloak around her, and looked about for a peg to hang it upon.

"I do not care if you walked or flew. The city is full of thieves and worse, Rosy." He took her cloak. "What you did was very risky."

"I must tell you something that just cannot wait."

He had just settled into Winter quarters for another few months of frosts and boredom. He could not think of anything that would be so critical that it could not wait until morning.

She looked at him with a terrible intensity. She walked slowly to him, lay her head upon his chest, and wrapped her arms tightly around him.

Howell returned the embrace. "Rosy, what is it that has made you risk yourself to come to me?"

She seemed unable to speak, but then found her voice. "Mr. Bell translated a conversation between Sir Horace and Colonel Mansfeld in which they concluded an arrangement to have you killed."

Howell could barely suppress his fury. Over and over this man had tried to kill him. Howell needed to end this. He needed to conclude his vendetta. He needed for Sir Whore's Ass to die.

"Bell said that Horace could not recall your name properly, but he had to admit to Horace that you were in Dichburg's Company."

Howell's first reaction to what Bell had done was astonishment that he had been betrayed by a man whose life he had saved at Zablat. Then he realized that it was the right thing to do. Mansfeld and his Captain would discuss this exchange at some point, and they would realize that he was Horace's intended victim. If Bell had not told them, he would look the fool, or worse, disloyal, and Howell would ultimately be served just the same. With this warning, Bell was doing everything in his power to repay Howell for his rescue at Zablat.

"If the Captain is going to have me killed, it will be Vlok who does the bloody work. I would kill that bastard this very evening, except that it would expose Bell. He has done me a great service, and I must do nothing to betray his trust."

She released him, and slid her hands down onto his as she stepped back. She gazed into his eyes. "There is more. You remember that I told you that Horace has lied to King Fredrick? He has persuaded Fredrick to believe that England will support him and the Rebellion."

"And that he gave the useless muskets to Fredrick to make his lie look as though it had some real substance."

"Exactly. All of this has made Fredrick believe that Horace is his greatest friend in the world." It seemed that her emotion was ebbing, and that she was returning to the keen mind that he knew so well. She dropped his hands, and sat upon the bench. "He's using that misplaced trust to further a plot he's made with the Emperor."

"Plot?" Howell sat next to the stack of firewood on the hearth in order to more easily keep his eyes on Rosy.

"Mr. Bell has gotten a peek at some papers that Sir Horace has in a portfolio that he carries at all times, and these papers discuss an arrangement for Horace to kill or kidnap the whole Royal family."

Fredrick, Elizabeth, and their children were the symbol of those who were fighting the Emperor. If he could find a way to eliminate them, it would go a long way towards suppressing the Rebellion. Sir Horace was more than willing to help the Emperor, of course. He would certainly gain a rich reward for yet another betrayal.

"It is all of a piece, dear. You know that he arranged the assassination attempt in order to disrupt the Rebellion, and insert himself where he could gain the most power. He knows that you hate him for what he did to you in Genoa. How could you forget that he got you thrown into that that horrible prison? So he wants you dead." She stopped. Howell could see that behind her resolute exterior, Rosy was fighting a battle with her fear that she would lose the man she loved.

"It makes perfect sense to Sir Whore's Ass to kill me now. He takes care of two problems with one stroke. He eliminates someone who might complicate his plan to kidnap the King and, at the same time, he removes someone who hates him enough to tear his head off with his bare hands."

Rosy forced a smile. "Fortunately, I am with Elizabeth in the Royal household, so I see much of what Sir Horace does. Between she and I, we may be able to prevent anything that he attempts to do." She paused, and looked at him with the same emotion on her face that she had had when she first came into the house.

They sat in silence, each regarding the other, each feeling the presence of the other.

"This may be our last time together," Howell said. Horace was going to try to kill him sometime soon. Horace would make a move against King Fredrick soon, and Rosy might die in her effort to stop him. "We may never meet again."

Plainly, she had considered this as well. She sat, looking at him and thinking. There was a look of determination, of courage in that face.

"You know, Rosy, we are a pretty tough pair to kill. He's thought to do it several times, and yet we've always won."

She stood, and he did too. She wrapped her arms around him again. She hugged him, but not so fiercely as before. She pressed her face into his chest. "We are one, dear. If you die, so do I."

He enveloped her with his arms, and murmured into her hair, "Then let's not do it."

She gave a little chuckle. She pulled back to look up at him with a little smile on her face. She wiped away some tears upon her cheeks. "Well then, what should we do?"

"I have an idea." He kissed her firmly, and disentangled himself. He stripped off his doublet.

She watched him for a moment, and then pulled out the bow at her neck that held her dress closed..

# Chapter XXII

## Howell Davidson is Murdered

Howell walked a cold and deserted road with three other men, the snow squeaking beneath their shoes. They had been at it since dawn and now the weak Winter Sun in front of him was as high as it ever got at this time of year.

*Achoo!* One of the new men in the foraging party had been doing that all the morning, and the snot soaking his beard was very off-putting.

"You there!" Vlok looked at the offender. "Quiet."

Theirs was an unusually small party, just two raw recruits shivering in their cloaks and the repulsive Corporal Vlok leading the way in his absurd coat. No, Howell corrected himself, this garment was not absurd; it was rather practical, really. These strange Russians or Hungarians or whatever you called them had a good idea when you came to think upon it. Vlok's coat amounted to a heavy doublet with skirts that hung all the way down below the knees. You got the warmth of a cloak and still had your hands free. He would probably enjoy one himself if he did not associate it with the vile Corporal.

The little band came to a side road that was indistinguishable from countless others that they had passed this morning, but this time Vlok looked back over his shoulder at them. "This will do."

When Howell had been ordered on this expedition, he had been told to leave behind all his musketeer kit. Vlok's order could mean but one thing, Vlok was going to shitte on him one more time, and use him as the pack animal on the way back. By rights it should be one of these two new men. Neither of them knew which end of a musket the ball came out of. But then, the injustice just helped to make Vlok's point, didn't it?

Soon they came upon a rapidly flowing stream. They walked along this until they came to a tiny village. Like the rest of the countryside they had walked through, there was no sign of movement here. However, across the creek, there was smoke coming from one of the two chimneys of a house rather bigger than the norm. On Howell's side of the stream were the usual dilapidated little sheds and tumbledown out-buildings. Vlok led them into a barn with only half a roof. Once they were inside, they all gathered in one corner. The barn held a pen with two miserable sheep, and the damaged roof over their heads. The animals gave off a tiny amount heat, and the barn's walls blocked the wind. It was little enough relief, but welcome for all that.

*Achoo!*

Vlok looked back and forth at his little command. He seemed unsure of himself. "I am going off on a bit of a scout." His manner was very nervous. "I will be back presently."

This was very odd, and set off alarm bells in Howell's head. He watched as Vlok moved across a little bridge that carried him over the stream, and led to the great house. This was not a military reconnaissance. Vlok was going directly to that house for a very particular reason. Howell did not know what that that reason was, but he strongly suspected that it had to do with the plan to kill him.

Howell turned, and scanned the woods behind them. Giovanni stepped out from behind a tree. So their plan had worked.

Giovanni had followed them without being seen, and now was in a position to cooperate as required.

"I am going to go take a piss," Howell said to the other two men.

*Uh hunh* Neither of them was in the least bit interested. They were far more interested in trying to keep warm.

Howell started off across the fields towards Giovanni. What, or more likely who, was in that big house? Perhaps it was Sir Whore's Ass himself. Should he go and find out? The snow he walked through was deep or shallow depending upon the level of the ground beneath it. If he went to the house, would he learn anything that would really help him? He rejected the idea. Unless everything went right, it would queer the whole idea of making him 'disappear' again. Horace would not search for a ghost. Howell would have a better chance to kill Horace if he thought that Howell was dead.

"Well, you certainly took your time."

Howell slipped around the big tree where Giovanni was sheltering. "Let's see if we cannot get Vlok hurrying to Hades."

"I thought you'd never ask."

"Follow me behind the side of the barn that's still in one piece." Howell pointed at the one intact side of the wrecked barn.

"The side that cannot be seen from the house."

"Exactly. When Vlok comes back, I will lure him back there, and we will beat him senseless."

"Oh, I am looking forward to this bit."

Before they could work out the details of their plan, though, Howell saw Vlok returning from his 'little scout'. They would just have to work it out as they went. "Follow my lead." The comrades went to the back of the barn, and Howell motioned to Giovanni that this was the place to await him. Howell stepped out from behind the barn, and into full view of Vlok and whoever might be in the house.

"You there!" Vlok stormed towards Howell. "What are you about? I told you to stay in the barn."

Howell hitched up his pants, and closed them up. He wandered casually behind the barn.

"You, Davidson! Get over here! Now." Vlok turned the corner to reach the back wall of the barn. He took one step. Then he saw Giovanni. He took another, slower step as confusion swept over him. *Unh*

Giovanni hit him square in his forehead with a stout branch he had brought with him from the forest. He rolled Vlok over, and was confronted with his coat. He had never seen the like. Howell unbuttoned it, and then Vlok's doublet. Giovanni was back in familiar territory. He plunged his knife into Vlok's heart several times. No one wanted Vlok to wake up, and become troublesome.

Now began a flurry of undressing and dressing that would ensure success of their deception.

They took off Vlok's coat and doublet.

Howell stripped off the distinctive red doublet that Rosy had given to him, and they put this on Vlok. Although it would probably never be seen, Howell wanted to make his deception as convincing as it could be. He took of his hat with its unique clipped brim, and put that on Vlok as well.

Giovanni took off his own doublet, and put on Vlok's doublet and then his odd coat.

Now Vlok was dressed as Howell, Giovanni was dressed as Vlok, and Howell was barely dressed at all.

"Excuse me, Sir. But can I ask what is happening?" It was one of the men from inside the barn calling over the wall. They had heard Vlok yelling his head off, and suddenly they could hear nothing but dead silence. The only surprising thing about this question was how long it had taken for them to ask.

"Cannot a man take a shitte without being bothered?"

Howell's answer satisfied the men on the other side of the wall, so he and Giovanni could complete their preparations. Howell helped get Vlok onto Giovanni's back. Howell whispered into Giovanni's ear. "Into the stream." It wouldn't be easy to fish the body out of the stream and discover it was not him, but he wanted to be as certain as he could be.

Giovanni struggled across the field with Vlok weighing him down. It looked for all the world like Vlok carrying Howell. They were visible from almost anywhere. Giovanni carried him to a place along the bank of the creek, and then he decided that enough was enough. He shucked off his hated burden, and let the unconscious man drop face-first, half in the snow and half in the creek. Giovanni adjusted his (well, actually Vlok's) coat. The brilliant snow behind him, lit by the low Sun, did a good job of disguising him, but still he kept his hat brim down so no one could see his face.

Vlok had shot him in the storming of Pilsen, but had missed killing him. Giovanni would return the insult, but would get the ending right. "You cowardly, filthy son of a whore. You would kill me?" Giovanni drew his short musketeer's sword, and held it dramatically over his head for an instant. He plunged the blade into Vlok's back. "You would kill my amico?" He wrenched the sword free, and jabbed it in again. "You miserable hog!" He delivered a third strike. "Die, rat! Die!"

Howell saw him linger over the prostrate figure. Then Giovanni shook himself, and delivered a mighty kick into Vlok's ribs. The now definitely ex-Corporal rolled down the bank, crumbled the thin ice lining the stream, and began to drift away. Giovanni had his revenge. He stood over Vlok, his bloody sword in his hand. He had satisfied his vendetta.

*Achoo!*

Giovanni looked back towards the barn, and Howell looked out from his hiding place.

The other two members of the patrol were standing, and looking at someone they thought was their leader, Vlok. They must have seen the murder take place. Their leader had just killed one of their number, seemingly without cause. Would he kill one of them next? Each of them pulled his sword.

Howell exposed himself by jumping out from behind the back of the barn. "Run!" Howell started towards the trees. "Run! Follow me!" He made his way through deeper and shallower patches of snow as he made for the trees. "Hurry." The other two men tried to match his pace, but with the irregular snow and burdened with their muskets and all their other kit, they quickly fell behind.

Howell glanced about. There was no sign of the two musketeers who were behind him. He was alone. He had neither hat nor cloak, not even his doublet. It was a freezing cold Winter's day and he was in his shirt. He kept up a brisk pace as much to keep himself warm as to distance himself from any pursuit. He plunged ever deeper into the woods. Giovanni had exposed himself at Genoa's gate to save Howell, and now it was Howell's turn to save his friend.

The woods screened him from view for a time, but it was not a big forest. He came to a field and paused. Looking behind him, he watched and listened for a long, cold moment, but there was no sound nor movement to reveal that the two musketeers were following him. Maybe they had lost the trail completely. Maybe they had just said *'To Hell with it,'* and started for Pilsen, or run off for who knew where.

Howell started across the field. He had not realized it at the time, but the trees had broken the strength of the wind. Now that he was in the open, he was chilling rapidly. He needed shelter soon.

Beyond this field was another margin of trees and brush. He penetrated it only to find another snow-covered field with more trees in the distance. Howell could see no sign of a building anywhere. He shivered almost constantly now. Aimlessly, he turned to walk along his current tree line, vaguely hoping that it might deflect some of the gusty wind.

Alone again. Giovanni had done what he could for him, but the result was what it had so often been, Howell Davidson versus the world. He would succeed or fail, Hell, he would live or die, based solely upon the choices he made. It was the world in which he had grown up, and he felt strangely familiar, even comfortable, with its dimensions. It was as though all his experience with comradeship and working with others had been a species of illusion. He had reemerged as a solitary animal, and a dangerously cold one, at that.

He came to the corner of the field where his border intersected another, and looked hopefully about. Diagonally across an open space before him he saw what looked like a denser wood. He immediately set off in that direction. He felt his exposure, the wind biting at him, the cold more penetrating. He was beginning to stumble in his haste to cross the field. The cold gnawed at him relentlessly. By the time he reached the trees, he knew he was in serious trouble.

Now he was in a substantial wood with trees stretching in all directions with undergrowth scattered throughout. He weaved among the various obstacles with no real sense of where he was going. Just continuing to move was success.

Then he caught a scent of wood smoke. Hope sprang to life, and Howell willed himself to stop, and try to determine its source. Standing still, even briefly, was agony. For one long, shivering, moment he could smell nothing, but then there was a breath of wind, and he caught the odor once more.

Instantly he was in motion, forcing his stiffening limbs to keep moving. He came to what appeared to be a solid wall of brush. He had no time to search for a way through, so he waded directly into the obstacle. The tangle of branches and vines ripped at his shirt, and clutched at his every step. He staggered this way and that, trying to find a way through. Ultimately, he fell, and lay among the stiff boughs.

He was not the least bit comfortable, with branches poking him in various unpleasant places, but oddly enough he felt like resting right there. He felt an almost irresistible desire to take a little nap, to close his eyes, and drift away into oblivion. Somewhere, deep in a corner of his mind, however, another voice was shouting at him. *'Get up, you idiot! Get up! Get up! Get up!'* He really did not want to hear it. *'Get up you moron! You'll freeze to death! Up! Up! Up!'* It was something like the two personas he could assume in battle. Here, one urged him to stay, and the other to go. He made an effort to seriously consider his options, and he made his choice. With great reluctance, he began to move, and suddenly reality swept over him again. He shivered so violently that he was paralyzed. When the shaking finally receded, he struggled to his feet. Slowly, he tried to continue through the brambles. Each step was a triumph of will power. Alright, take another step. Now another.

Suddenly he caught the smell of wood smoke again, powerful, like it was near at hand. It filled him with indescribable joy and crazy determination. With a tremendous effort, he broke free of the bushes and stood in open ground. Deliriously, he looked about, and saw a small valley before him. His salvation seemed to be down there somewhere. He followed the scent of smoke down the slope, and then he saw a small fire barely visible through the trees. It flickered, it beckoned, it seemed to him like a tiny opening into a barely imagined world of warmth and safety. Stumbling, at his last stretch, Howell staggered into a tiny circle of dirt surrounding a tiny fire. He felt himself losing consciousness, and toppling into the flames.

# Chapter XXIII

## Unexpected Meetings

Howell awoke. He was sitting with his feet near a camp fire. His toes were warm enough now that they hurt, which Howell hoped was an improvement over their previous numbness. He had some kind of cloth tied around his head, and was enveloped in blankets that smelt strongly of domestic animals. It was not disagreeable, and he found that he was comfortably warm. It was still daylight, so he could not have slept for long. He roused himself, and took in his surroundings.

"Ah, waking up are we?" The words were in French.

Howell banished the last of his sleepiness, and looked towards the voice. A man sat close to the fire, enveloped in his cloak except for a protruding hand holding a quill. There was a book open upon his lap. He was probably slightly older than Howell, and under his broad-brimmed hat he had that long nose and thrusting jaw that some people think is typically French. "You invited yourself to dinner in quite a dramatic fashion yesterday, my friend."

Yesterday? Had he really been so far gone that he had slept for most of a day?

"How do you fare?" The man had now switched to French-accented German, doubtless assuming that anyone who came blundering most-frozen out of the woods would be more likely to speak German, or Czech, rather than French.

How should Howell respond? Who was this man?

Now another fellow stepped into the circle around the fire. He was dressed like the first, with his hat pulled down low and a length of cloth wrapped around his neck. His cloak was almost long enough to drag along the ground. "Has he said who he is, Monsieur?"

Apparently, French was the language du jour. Howell was more likely to get sympathy and help from this pair if they thought he was one of them, so French it was. He would never pass himself off as a native, though, so he had be Belgian or Alsatian or somesuch. How about an identity that was simple, believable, and utterly forgettable? How about the most common name in use in France? "My name is Martin," he offered in French, "Jean Martin."

The man seated near him leaned across his book, and extended his hand to Howell. "I am Rene Descartes, currently amongst the officers serving his most Catholic Majesty, the Emperor."

Howell made a bit of a show of struggling to get his hand free while he considered what he should do. There were plenty of Catholic Belgians in the Imperial armies, but would this officer welcome an apparent deserter? Howell did not know about Alsatians, so he decided to just leave it vague. Howell at last freed himself from his swaddling, and took the offered hand. "Pleased to meet you, Sir. Thank you very much for your help." He hoped that his obscure answer would make it possible to later claim to have deserted from either side, or to have never served at all, depending upon need.

Descartes' traveling companion now came around the fire, and handed Howell a large chunk of bread. He had not realized how hungry he was until he took the first morsel into his mouth. The march yesterday morning, his escape, and now hours of sleep, all in the punishing cold, had left him ready to eat the whole loaf, and look for more. "Merci," Howell said to the servant. He tore into the bread greedily. He turned once more to Descartes. If he could engage this man in conversation, he might learn something that would help him later. "Please, Sir, what is that you are writing?"

"This?" Descartes raised his hands in despair over his note book. "This is the only result of years of struggling with a grand, majestic, and damnably elusive problem. And after all this work, I have finally seen that it all depends on first principles. That is the crux of the problem. What can one truly believe?"

Howell truly believed that he needed to be done with this odd man, and be on the road again. Once he had swallowed, and started on the next mouthful, he said around the bread he was chewing, "I wonder if I might impose on your hospitality again, Sir. Your horse blankets have served me very well, but I expect that you will want them so that you may continue your journey. Do you have a cloak, or something of that sort, that you could spare?"

*Hm? Ah!* "Of course. I beg your pardon." At this point Descartes had so far abandoned his philosophical musings that he was capable of noticing that Howell had no cloak, only saddle blankets wrapped around his shoulders. "Alphonse, give Monsieur Martin that," *ahhh…*

"Coat, Monsieur?

"Yes, yes. That strange new garment I was given."

"Immediatement, Monsieur."

There was a pause while Alphonse worried the baggage, and Howell squatted to take full advantage of the fire. Descartes sat staring off into his own version of the universe. The servant returned with another of those odd coats like the one that Vlok had been wearing. The servant helped Howell get out of his temporary wrappings, and into the coat. It was not immediately warm, it had just been unpacked after all, but it covered him completely, reaching below his knees. Although it was of the same style as Vlok's coat, Howell now viewed it in a very different light. It was just the thing for a spot of traveling. He looked up, and found that the servant had read his mind, and was offering him the rest of the loaf. Howell stuffed the half-loaf into one of the pockets so that he had have something for later. He was not sure if the coat was a gift or a loan, so he simply said, "You are most kind, Sir."

"Certainly, certainly," mumbled Descartes, his mind drifting off again into the philosophical maelstrom that only he could sail. "I would happily give a hundred such garments for the insight that eludes me."

Howell needed to make a start on his journey to find Rosy, and it could not be soon enough. He would have to leave this Descartes fellow to work on his incomprehensible task, while he got on the road. "I beg your pardon, Monsieur, but I believe that I must excuse myself for a moment."

*Hmm?*

The servant understood what Howell meant and said, "The gentleman needs some time alone."

"Alone?" Descartes seemed oblivious of the conversation going on around him, but suddenly stood and his eyes burned. "I am consumed with the need for certainty!"

Consumed by something, and no mistake.

"I have cajoled and flogged my mind for a way as precise as mathematics to encompass the world around us in all its complexity, but before I can delve into these mysteries, I must have an unassailable foundation! Without that, how can I truly understand motion, or growth, or how it was that the Earth came to be?"

What Howell needed was to escape from this babbling lunatic straight away. He blurted out the first thing that came into his head in answer to this man's crazy questions. "You know, when I am thinking what I am thinking, I think, well, someone's got to think it, do not you think?"

Descartes fell silent and got that far-off look in his eyes again.

*'Good,'* thought Howell. *'That's shut him up.'* Howell began to ease himself away from the campsite.

The Frenchman's little smile began to broaden, and Howell heard the last of the man's ravings as he made his exit. "I am thinking, therefore I must exist." Descartes clasped his book to his chest, and stared off into space. "I do not see any flaw in that logic. It is true. It *must* be true." There was something like ecstasy in his voice. "It is the First Principle that I have sought all my life! I think, therefore I am! Cognito ergo sum! It is so obvious! How was it that I could not see it until now? It's brilliant!"

Howell mumbled something that no one listened to, and made his way quickly back into the trees. Rather than evacuating his bowels, however, he evacuated the area.

He quickly came upon a main road, and began to follow it to the east. The weather was still as cold as it had been when he left the camp of the raving philosopher, but the coat was doing its job well. He had a half a loaf in his pocket. He had a daunting journey ahead of him, though. Were the coat and bread enough to see him through to Rosy?

His immediate problem was needing to know if he was on the right road. The way he trod led east, but would it take him to Prague? So far, he had not come to a meeting with any other major thoroughfare where he had have to make a choice, but how long would this last?

He saw a village. It was not one of those that were right along the way, they would all have been thoroughly ransacked long ago, but this one was off in the distance. Howell turned into the lane to see if he could find something of value. It was well past harvest season, but Howell could see no evidence that there had been crops in the fields around him anyway. None of the buildings in the village had a roof. Not only was the thatch all burnt away, but even the rafters where black. That did not mean there was not a hat, or a strip of cloth to wrap around his ears, or something else there that might help a cold man.

Then he was struck by something he saw out of the corner of his eye. There was a man upon the road, which was no surprise, but this fellow had turned into the same little lane as Howell had. This was an odd coincidence. He was not making much of an attempt to hide his approach, either.

Howell decided that caution was his best course. He walked casually into the first building he came to, stepped to the side of the door, and put his back to the wall. He drew his knife. He waited. Might this man have been sent by Sir Horace? Supposedly Sir Whore's Ass thought him dead, but had something gone wrong? Had someone seen him escape?

The man stepped into the room. Instantly, Howell had his arm around the man's neck and the tip of his knife against the man's ribs. "Who are you?" he demanded.

The man laughed. "I am King bloody Solomon, you idioto!"

"Giovanni!"

Howell's amico turned around and the two hugged one another like lost brothers.

"This is amazing!" Howell held Giovanni at arm's length, admiring what seemed like an apparition. "How did you manage to find me?"

"Well, as soon as I saw you disappear into those trees, I started trying to work out what you'd do next. So I am thinking that an ambitious, strapping fellow like yourself, out on his own…"

"…freezing his balls off…"

"…what's he going to do? Where is he going to go? Horace has pledged to the Emperor that he's going to kidnap, or kill the Royal family. I thought upon this. What does this mean for my amico? Finally it came to me. It meant that you had to come to this very spot."

"What?"

"It is obvious. If Horace is going to take Elizabeth to the Emperor, then your Rosy stands in his way. You cannot let that happen. You *will* not let that happen, so you must take the shortest time possible, and that means taking this way. I have wandered around this stretch of road ever since you were murdered, expecting that we'd eventually fall in with one another."

"That's dead brilliant. I knew I made a good choice when I picked you as my friend. But as you say, I am in haste." Howell now searched the building in which they were, and then moved from one house to the next in the ruined village. Giovanni searched others. Howell found the hat that he had hoped for, and Giovanni some lengths of cloth. Howell wrapped these strips over the top of his head, and tied them under this chin to keep his ears warm. He put on his 'new' hat. "Let us be off."

They went down the lane, and turned onto the road that Howell had been walking. "Do you know if this is the road to Prague?"

"No," Giovanni said flatly. "I have not done much but wait for you." He displayed a sack that hung inside his cloak. "I did learn where there was a little left to eat as well."

"Good." Howell pulled out his bread. "And I have a half a loaf here."

"I thought I was the bread thief," said Giovanni.

"When we were looking for a way out of Genoa, yes, but now I am the robber," said Howell.

"It was not robbery that you were about yesterday." Giovanni got quiet. "Drawing off those musketeers yesterday was a very brave thing you did for me."

"And standing up to Sir Horace at Genoa was a very brave thing you did for me."

Giovanni was back to his familiar self. "Right. We've each had our turn at this bravery thing, so we won't have to do anything like that again."

Howell smiled, and slapped his friend on the back. "I'm sure you're right."

Their very long trip resumed.

Every now and then a wagon passed. For a while Howell tried to wave them down to get a ride, but eventually he gave up the attempt. There were many people on the road, and it would be pure luck if a wagon picked them out from the mob of refugees. These were the cast-offs from the fighting. The poor wretches who had no idea what had happened to them, nor why. Some plodded along, with no idea of where they were going. Some begged. Some lay motionless on the margins of the road, or even in the midst of it.

He should care about these people. He knew that each one of them needed help. No one would expect him to save them all, but he knew that using his limited resources to help even one of them would be an honorable act. It was hard, but he knew that it would be none of these people that he would help. He must conserve whatever he had to help Rosy defeat Sir Horace. He looked at Giovanni.

Silently, Giovanni returned the look.

"I am going to stand by her side."

Giovanni offered the slightest smile. "Somehow I knew that you were going to say that."

They walked while there was day light, and slept in whatever corners of whatever buildings they chanced upon as each Sundown overtook them. Nothing but waste and misery remained in Bohemia, so they had no choice but to rely upon the food that each of them had brought with him.

As he tramped along, Howell tried to work out what lay before him. He was making his best time for Prague, but he was not really sure if Rosy was there or not. He had to assume that Rosy would stay close to Elizabeth, so if he could find the Queen, he would find Rosy, but he could not be certain where that was. Perhaps the Royals, with Rosy at Elizabeth's side, were still in Pilsen. Were they on their way to Prague? Had they already arrived in the capital?

Horace wanted the Royals, dead or alive. Rosy was determined that he should not succeed. What did Rosy think she needed to do to stop Sir Horace? What had she already done? What had already happened with Rosy, the Royals, and Sir Whore's Ass? Sir Horace may have already sprung his trap. Rosy would have done all in her power to stop him, including sacrificing her life.

Howell put it out of his mind. Rosy was in Prague, and he would get to her in time. He would kill that miserable Sir Whore's Ass into the bargain. That was final. The details would sort themselves out as he went along, but he knew how it would end.

There was, however, one thing that was in his power that could help the situation. "Let us pick up the pace, Giovanni." Howell did what he could to push his tired, cold, hungry body to just a bit more speed.

"You know amico, no matter how hard you push me, I am never going to fly like a bird."

Howell tried to coax a laugh at his friend's jibe, but it just wouldn't come. He could not help himself. He could not help but focus on his speculations about Rosy. He could not even think about the world around him. It was enough that he kept his legs moving.

"Warm Sunshine and nice puffy clouds are so boring. Thank God the weather is getting better."

*Hmm?*

Howell had not noticed it, but it was starting to snow. The wind was picking up. Very soon, it was driving sleet into the sides of their faces. The scrap of cloth that Howell had around his ears was quickly soaked. The road could not decide if it was ice or mud, but it provided treacherous footing either way. Howell's desperate race to Prague was now proceeding at a stumbling pace.

Then Howell heard the jingling of horse harness and the sound of multiple horses' hooves in the slop behind him. He turned to see a coach emerging from the cutting sleet. The driver had his head down against the wind, and his reins were slack. He was letting the horses make the best time they could through the terrible conditions. This was all innocent enough except that the coach was coming directly at Howell and Giovanni.

"Hold there!" Howell shouted through the snow at the driver. He grabbed Giovanni, and dragged him towards the side of the road. "Beware, Sir!"

The driver's head snapped up, looking for whoever had called to him. He made no effort to rein in his horses, though. He could not pick out Howell and Giovanni. He called out blindly into the driving snow, "Clear the way."

Howell and Giovanni stumbled onto the sleet-encrusted grass that lined the road. As they regained their balance, the coach rolled closely past them. Howell's legs and the hem of his coat were doused in mud.

"Thank you for your consideration," Giovanni yelled up at the coachman.

The driver finally saw them, but all he did was give Giovanni a blank glance.

Giovanni returned the insult by flicking his fingers off the bottom of his chin.

Howell's reaction was entirely different. This coach was almost certainly headed for Prague. He was bound to come to a fork in the road, and if he could follow this vehicle it would lead him in the right direction. He took off at the best speed he could manage in the treacherous footing. He listened at the jingling of the harness as along as he could, but eventually it faded away. It seemed that the last thing he heard was ever so slightly to his left.

He did come to a fork in the road eventually. He took the left branch.

Time and distance, dangers on the road, hunger, and the very weather itself were against him, but Howell would not fail..

# Chapter XXIV

## The Army of the Damned

It was freezing. Their breath puffed out before them as they shivered. The cold gnawed at them as they made their way along the road. The wind came and went in unpredictable gusts that bit into Howell in the areas not covered by the coat that he had had from that gibbering Frenchman. Their stomachs growled. The road had been mercilessly churned when it was largely mud, and now it was frozen into irregular ruts and unpredictable mounds, so that he and Giovanni made slow progress. They had not eaten yesterday, and they had not eaten but little morsels in the days before that. Howell was feeling it, but he had to get to Prague, to his woman. Who knew what devilment was going on while Howell struggled to get there? He had to keep moving. He had to move faster than this.

"Is that smoke?" Giovanni pointed ahead.

Howell looked up. It did seem that there was a smudge beyond the line of trees in front of them. The more often he looked up from the treacherous road, the more convinced he became that he saw the smoke of many fires. Fires where he could warm his hands. Fires where food was cooked. "We will just stop long enough to get a mug of soup, and then we will be on our way." Howell knew that it was not enough, but he just could not bring himself to tarry any longer than that. He just had to get to Rosy.

"I was thinking more in terms of a roasted chicken with berries and…"

"They'll probably have that tomorrow."

Giovanni's good humor was just the same as it always was. He was ever the good comrade. He was just as worn and hungry as Howell, but he did not complain.

Despite the terrible footing, the friends hurried forwards. With each step the smoke was more easily seen. Now it was smeared across a wide swath of the sky. Now Howell could smell wood smoke. This must be coming from many fires. Were these camp fires, or the houses in a burning village? He had certainly seen more than his share of those in these three years of war. He did not have any choice but to plunge into whatever it was that lay ahead, and hope that it was a place of temporary refuge.

They came through the trees, and saw two or three camp fires. No, more. Howell and Giovanni jostled into one another in their haste. There were many fires ahead of them. There were fires dotted everywhere on the ground ahead of them. They helped one another from falling down. Howell could see fires everywhere with men sitting around them warming themselves.

In fact, there were so many men that Howell knew this must be an army. But which army? Plunging into this camp was a chance that he had to take. He had to reach Prague in the shortest time possible, and he could not stand beside Rosy if he starved to death. Howell had to continue no matter which army this was. The same was not true of his friend, however. "Giovanni, this may be the Army of the Bohemian Estates, or it may be one of the Emperor's armies that are seeking to put down the Rebellion. I have no choice but to try to get something to eat here. You may choose to do something else."

*Ahh* "Let me see if I understand my choices correctly. Down one path is certain starvation. Down the other is a chance that I will live." *Hmmm* Giovanni led off. "You have persuaded me to follow you."

Howell and Giovanni went into the encampment. Men were sitting in front of a fire, or walking around everywhere they looked. However, Howell saw muskets and spears lying haphazardly on the ground. He had seen enough of soldiering to know that this army was in terrible straights. The men he and Giovanni passed were poorly dressed, but there were no sutlers' wagons in the camp. Everyone looked as though he could not remember when he had last had a good meal, but Howell saw only a few men tipping back a mug, and no one who was chewing bread.

Giovanni approached one of the men sitting near a poor fire with a crude pot hanging over it. "Suppe?" he asked, showing their two mugs. The man turned, and looked dumbly at Giovanni's outstretched hand holding the mugs. Then he turned away. The friends moved on to the next fire. They tried fire after fire amongst the ragged host. Nowhere did they have any luck.

Now a man of very different appearance approached them. He was better dressed than anyone else that they had as yet encountered. This man carried a sword, and it was not the sort of crude blade that was issued to musketeers, but rather something that looked to be of considerably better quality. "I am Captain Erlach," he said as he arrived. He seemed wary about these two who were wandering about in his camp. "To whom do I have the honor of addressing?"

"Well met, Captain." Howell touched the brim of his battered hat. "My name is Howell Davidson."

The Captain extended his hand. "If I may ask Sir, what is your business here in my camp?"

Now Howell was up against it. Which army was this? If he claimed to be a rebel, and this was an army of the Emperor, he would have to escape as best he could. He looked around at the sickly wretches that surrounded him, and decided he could outrun them.

"My friend and I were hoping that you could give us some food so that we may get to Prague."

"And what will you do in Prague?"

Howell gave an inner sigh of relief. These were soldiers who fought for the Rebellion. "I will reach a woman there who serves the Queen. I wish to take her from such an exposed position to a place of safety."

"The King and Queen are in the castle, but I fear that you may have difficulty gaining entrance."

Now Howell could be sure where Rosy was. He had been right to push himself and Giovanni as hard as possible towards Prague. "I am sure that I will find a way," Howell said with determination.

The officer stood quietly thinking. Then he made a decision. "Their Majesties cannot know the state of their army, else they would make great efforts to correct the problems you see here." He swept his hand around the miserable camp. "Perhaps you will tell your woman about what you see here, and that she will be able to get a message to their Majesties." He had begun speaking with exhaustion in his voice, but now he sounded more hopeful. "Give them some soup," he said to one of his men.

Something like ecstasy swept over Howell as he swallowed what he was given. Food. Solid evidence of where to find Rosy. A message to deliver that might help him gain entrance to the castle. He drank his soup bursting with confidence.

"What you see around you are the men of my Company, Mr. Davidson, and they need food just as much as you. Our Commander in Chief, Sir Horace, does what he can, but we have too little shelter, weapons, and too few horses. We have almost no gun powder at all." Erlach was making an effort to put right what was plainly wrong with his Company, but he was taking a terrible chance by telling all this to a man that he had only just met. "My men need clothes. There is disease in the camp. Some who are now in this camp will die. Many already have."

What could Howell say? That it was all the fault of Sir Horace who had convinced King Fredrick that his muskets were all that the Army needed?

"Bad as all that is, we have managed to hold on because of ordinary soldiers who encourage their comrades. Men who put up with these terrible conditions so that their country may be free. My men understand what this army is fighting for, and will stand with their mates no matter what the hardships."

Howell had never considered the cause of Bohemian independence in this light. Now he saw a whole country fighting for what he had always fought for. They wanted respect, but not as individuals. They wanted respect as a nation. Suddenly he cared about what happened to Fredrick and Elizabeth, just as Rosy did. He was no longer searching for the Royals just as a way to find Rosy. He wanted the King and Queen to survive, so that no matter what happened, the people around him could look at their Majesties as the symbol of the pride they felt in their nation.

A commotion suddenly burst out in the camp. A man ran up to Erlach. "Something's going on, Sir. Cavalry are riding everywhere, and yelling like Hell."

"Here comes some horse now." Giovanni pointed out a small band of cavalry that was riding raggedly towards them. Their commander was yelling at them, and several of the men were struggling to catch up. "Mio Dio!" Even at this distance, it was obvious that they were at full speed.

Without a word, Captain Erlach and Howell set out at a dead run to cut them off. They both had the same thought, that they needed to know what was happening.

"Stop! Stop!" Erlach shouted to the fleeing horsemen as he ran towards them. "Stop! Report!"

Most of the troopers roared past before he and Howell could intercept them, but they managed to flag down two stragglers. Howell grabbed one man's bridle, and Erlach got a hold of the other one's stirrup. "Report!"

The two men were beside themselves with terror.

"Trooper! Report!"

The man that Erlach had stopped gathered himself, and looked down at him. "They are just over there, Sir." He pointed back the way he had come.

"Who's over there?"

"Who? The enemy, that's who."

"Horse or foote?"

"Both! They are everywhere."

"How many?"

"Thousands and thousands." The man glanced over his shoulder as though he expected to see the fields behind him covered with enemies at any instant.

"Calm down. Tell me exactly what you saw."

"I saw their whole damned army, that's what I saw!"

"Calmly, soldier. Calmly. You were riding north, and you came upon something unexpected."

"Yes, Sir." The man swallowed hard, and nodded. "We had barely gotten there, and we saw that the road was just packed with them. As far as you could see in both directions."

"And which direction were they headed, trooper?"

He raised his arm, and pointed towards Prague. "That way, Sir."

Erlach took this with no evident emotion, but if the enemy had stolen a march on them, and were already well on their way to the capital, things would go hard. "Did you see any guns?"

"Yes, Sir. They had a lot of cannon. There was a dozen or more of those big fuckers. And musketeers and spearmen and men on big black horses all covered in iron."

"They'll think we're deserters, Sir!" The man that Howell restrained looked anxiously to where his mates had disappeared only moments before. "Can we go?"

The Captain released his hold, and stepped back. "Good man. Rejoin your unit."

Howell freed his man, and cleared a path. The two cavalrymen dashed off towards the village. By this time, Giovanni and several others had joined their group.

Erlach was thinking hard. "If the Imperials have already gotten 'round one end of our line with an army big enough to include a large train of artillery, we're going to have to move quickly." He did not appear to be speaking to anyone, but rather thinking aloud. "They keep us amused with what seems like settling in their new Spanish allies. All the while they are moving their whole damned army around our flank where the clods we work for have no scouts." He glanced towards Prague. "We need troops back there to block them as quickly as may be." He needed his Company in formation to travel. Captain Erlach called out to those around him. "I want every man ready to march immediately!"

"Immediately! Yes, Sir!"

Howell and Giovanni appointed themselves Acting Corporals. There were no two ways about it. The Rebel army had to be in place before their enemies arrived. They moved through the camp, passing the word, getting the men on their feet, waking the sick and the terribly hungry. Getting this lot to start on an exhausting march, much less keeping them at it, was going to be a terrible challenge. They herded the men they collected throughout the camp, and added them to the column that was assembling for what everyone knew was going to be a Hellish march to Prague. The men who were forming up could only imagine what lay ahead of them. Howell knew that he was trying to get these men to where they would face men who were better fed, better armed, and in better health.

Captain Erlach anxiously watched his Company assembling. By now he was mounted, and rode here and there, straightening out a miscommunication or confirming a decision. His exhausted men formed up as quickly as they could.

A dispatch rider came thundering across the field. He rode directly to Captain Erlach and handed him a note. While the Captain read, the last of his troops fell into place.

"Bloody fools," he muttered to himself. Then he sat up straight in his saddle, and addressed his command with confidence. "Men! The enemy has gotten 'round our flank, and is even now marching upon our capital. We are in order, so we will not wait. We will get on the road, and lead the Army. We must make haste to put ourselves in front of the enemy so that we may get into a position to defend Prague. I want you to march like you've never marched before. Follow me." He turned his horse's head to the east, and started off. Those who could, fell in behind their Captain, and began their struggle to keep up. Other men from other companies, and even from other battalions joined them in the chaotic mob. Their marathon began.

Howell would join them, but first he wanted to be armed. He walked around the camp, looking for a musket. They were not hard to find. Any number of them seemed to have been abandoned by these wretched men. Howell found a bandolier that seemed to be in good condition, and looked in the chargers. Fewer than half of them were full. The Army had been starved of power by Sir Horace, and here were the fruits of that deception. Howell went from here to there, finding chargers in other bandoliers that contained some powder, and filling his own with their contents. He found a length of match. He also found the personal effects of the poor soldiers who had spent so much of their lives here. Many of them had left the last breath of their lives here as well.

Giovanni had equipped himself as well. Now they had everything that they were likely to get by way of support for the dash to Prague and to Rosy. "Let us be off, then." They would walk along the same road that these soldiers were using, but Howell was determined that they would make better time.

Howell inserted himself into the column, and began to pass one man after another as he sought to get to the front of the line and reach Prague as quickly as he could. Those he went around were tired and hungry. They had been campaigning without shelter for the whole of the Summer. They had been poorly fed, and not a one of them had gotten a change of clothes. Bad as that was, what lay before them would be the most difficult thing they had ever done in their lives.

As he moved forward, Howell found that the column was beginning to assume a sort of character. There were some who, for pride or out of a sense of duty, tried to maintain the pace that their Captain set, even though it was a struggle for them. There were men who knew at once that they would be unable to keep up, older men, injured men, sick men, and these sorted themselves out as they fell back to the best pace that they could manage. There were some who just quit altogether, and directly all these had been weeded out.

The men in the column were trying their best to do what was required of them, but it was not enough for Howell. He forced himself past these men to achieve his own goal. He pushed himself hard. "Excuse me, mate." He got around another man. "Coming through." Howell had to look upon these men in the same cold fashion as he had looked upon the refugees he had passed on the road. "Oh, sorry." He had to leave them to their fate. Howell would not let fate determine his future. Howell would make his.

"On your left." Could he make any real difference in what became of these men in any event?

"That's the spirit men!" The commander of the men trudging along the road walked his horse from the head of the column back towards Howell. He had an encouraging smile upon his face. "We're leading the Army to a place where we will prevent the enemy from taking Prague, and we're just the lads for the job." He urged them to what seemed a blistering pace to men who had scratched for their necessities for as far back as any of them could remember.

It was a cold day, but Howell was soon sweating. He ached. He had not had a good night's sleep since he had left Pilsen, unless one counted the night at the fire of that blithering Frenchman with his notebook. He continued to work his way towards the front of the column.

Erlach waved at Howell in a friendly way as he passed him on his way along the column, but his smile seemed to be a bit forced. Certainly he was worried about the coming battle, but was he also worried about the trust he had put in Howell?

It said a lot about this man that he was not yelling at his men to make way for him, or using his riding crop to beat a path, but rather he was asking firmly, but politely. "Excuse me, lads. Pardon me, I need to get through." His men raised no objection, and got to the sides of the road to make a path for his horse. He rode up next to some men who looked to be close to giving up. When he came up beside them, he chided them good-naturedly, and they responded in kind. While they were talking, they picked up their pace ever so slightly, and the change seemed to continue after the Captain moved on. One group at a time, one man at a time, Erlach put heart into his command. It was a grinding march, but he kept a surprisingly large portion of his men plugging along.

Howell's mind moved again onto something beyond the immediate rigors of the march. Somewhere ahead of him was his Rosy. Capable as she was, fending off Sir Whore's Ass or other threats might be beyond her capabilities. He needed to be with her. There were thousands of soldiers between him and her as well. If the Emperor's army and their Spanish allies reached Prague first, then they would block his access. Everything he wanted to do, everything he needed to do, would become impossible. The consequences of failure were just too terrible to imagine. If he was to get to Rosy in time, his best option now was to keep these men around him at their task.

Hour succeeded tedious hour. Mile followed frozen mile. Those who were going to give up had done so, and those with the strength to continue had settled into a steady pace. They were not flying, but they were getting closer to their objective with every step.

Howell was getting closer to his objective with every man he passed. He was as hungry as these men were. He was as exhausted as they were. But his commitment to his final goal was far greater than theirs. He pushed himself ahead as fast as his battered body would allow.

Suddenly, he was through. No one was on the road ahead of them. Somewhere ahead was Prague and his Rosy, and he had a clear path all the way to her. Now he could finally make the speed he had craved.

Howell found he and Giovanni were hurrying along the margin of a small river with a strip of trees and bushes separating their road from the water. Across the stream was a bluff that looked down upon them. Howell thought he could hear shouts of command, and then he was sure. "Giovanni, we have no wagons, yet I hear them." Howell looked across the little river, and saw something incredible. "Giovanni. Look across the river."

"Merde."

The opposite bank dropped gently to a small plain of sand and gravel just above water level. This was probably an area that was swept by the stream when it was in flood, but now it was just a wide stretch along the bank. There were enemy soldiers all over this gravel, drinking the water, and filling anything they had that would hold water. There were dozens of them, and their Corporals were yelling, and using their sticks upon them to get them moving again. The effect of all this was that the flow of their enemies was continuous, in a sputtering sort of way As soon as one squad had been put back on the march, another would come crowding down to the water, and the whole process would start once more. The men who were forced to move off the sand bar climbed up the slope to the top of the cliff. Along the top of the cliff, it was possible to make out men's legs walking, and horses, and wagon wheels.

Howell casually walked behind a small tree. "There." He nodded towards a tiny gap in the greenery on the opposite bank. Along the road along the top of the cliff there were several wagons, a huge gun being pulled by oxen, and foot soldiers mixed in with them everywhere. From the top of that bluff, they could easily look down upon the road that Howell and Giovanni traveled.

When the enemy discovered the column behind Howell, they would place their musketeers along the bank, and shoot at any of Howell's comrades who tried to pass. The Army of the Bohemian Estates would be disrupted, or even stopped, and the enemy would be able to seize Prague unopposed. He could not let the enemy know they were there, or there would be a fight. They just did not have the time for that.

The road here could not be used, yet the men behind Howell had to keep moving. Howell reversed course. He made his best time back to the head of the column. "Sorry lads. The road's blocked up there. We will just take this path." There was not any path. Howell was going to create one as he went along. "Here we go, lads." He struck off into the woods. "Just follow me, and we will be done in no time." As he began to lead his new column into the trees, he saw Giovanni take his place in the road, directing the men to follow him into the woods.

He needed to break a trail as quickly as possible in order to keep the soldiers moving, but the best he could manage was to weave back and forth between the bushes. Even then he found obstacles that he had to stamp down with his feet. It was exhausting work. He could not count the number of days he had walked one road or another. He was subsisting on some crumbs of bread he had had some days ago, and a mug of soup today.

How far should he take his detour? Surely he was well past the sand bank and the enemies drawing water there. Was he past the overhanging cliff? Was there yet another hidden danger that he should avoid? There was no way to know, but his trail through the forest had already cost the column precious time. He took the chance, and began to edge back towards the river. There was no sign of the bluff on opposite bank of the stream where he had seen the enemy. Howell led his column towards the river, and suddenly he was on the road.

Howell stood to one side. He waved the men who had followed him out onto the path. "Here you are, mates," he said. "Nice flat road, eh?" He tried to look cheerful and strong, but he was as exhausted as the men around him. He had gotten Erlach's Company around what might have a disaster. That was satisfying enough, but he realized that he had done something to help these men after all. He had made a contribution that he had never expected to make. The Captain had said that his men encouraged their comrades to endure terrible conditions. The men Howell saw marching by him weren't just the steadfast men that Erlach had shown him, but these were his comrades now. He smiled, and encouraged them as they emerged from the forest.

*Ahh* "Loafing again I see." Giovanni came out of the woods. "Captain Erlach assigned another man to my hard duties so that I might find you."

"Good." Howell shook himself to loosen his worn body. "Let us be off."

Once again the friends began to work their way up the column. As before, they continued to move just a bit faster than their comrades, and slowly make their way towards the front of the men who were enduring the endless, mindless trek.

It went on hour after hour. The afternoon wore on towards evening. The men around them were at the limits of their strength, going through the motions of walking while they tried not to think about their sore feet and their sore backs. Worse, everyone knew that there would be no supper tonight.

Howell in particular wanted to know what loomed ahead. He needed to know what he had to overcome to reach his goal.

Now there was a commotion at the limits of Howell's vision along the winding road ahead. A party of horsemen was moving towards him, against the flow of exhausted men. The setting Sun lit the backs of the plodding soldiers, and fully illuminated the faces of the approaching riders. Howell was not sure about the identity of the first man in this group, but he certainly knew the second.

The blonde curls could leave no doubt. He was the army's Commander-in-Chief, Sir Horace him-sodding-self.

"God's nails!"

"Mio Dio."

Unbidden, an image raced through Howell's mind in which he knocked Horace off his horse, and watched as he was trampled to death in the middle of the road by the clutch of toadies riding around him. What would really happen, of course, was that he would disguise himself, and hold his need for revenge tightly against his breast. Still, this latest fantasy wouldn't fade.

"Another blessing from Saint Justin, no?"

Howell wanted a blessing that ended with Saint Justin giving last rites to Sir Whore's Ass. There was another thought that crept into Howell's mind at this moment as well. Giovanni was always insisting that Saint Justin was looking out for him. Against all odds, he kept meeting Sir Horace. He kept seeing him in different situations. Perhaps there really was something to this Saint Justin business.

Howell changed his mind about donning a disguise. He wouldn't hide from this man. His hat and the rag that was already tied around his head would be all the concealment he would use. Howell would not be foolish enough to cry out or attack Sir Whore's Ass, but he wouldn't shrink from him either. This was his mortal enemy, the man who threatened his woman, the target of Howell's vendetta, the man who sought the ruin of the men who marched around him. Howell would look this cur in the eye.

The leader of the procession that was coming towards them was in full voice, talking continuously as he bulled his way down the marching column.

"Soldaten!" The orator was speaking in German to these men who mostly knew only Czech. "Soldiers! Your day of glory approaches!"

Several men in the Company had heard Howell speaking German, so the men nearest him crowded around. "What does he say?"

"Good day. Big day," Howell said in his poor Czech, but he still focused on Sir Horace.

"I'd rather have a meat and cheese day." "And beer."

"Soldiers! Patriots!" The rider swung his sword over his head in a vigorous display. He was wearing a heavy bearskin cloak held closed with shining brass hooks. After each proclamation, he wiped his nose on the sleeve.

Sir Horace approached. Howell raised his eyes, and beheld the face from his nightmares. This was a face that he had not seen since that Hellish night that they escaped from the burning village of Zablat. This man had wounded him by running his rapier through Howell's shoulder, although he had been aiming at Howell's heart. Howell looked squarely at that face without fear. He would kill this man. He would do it because Rosy needed him to do it. He would kill this man to fulfill his own need as well. There was not the slightest doubt in his mind.

Sir Horace looked neither left nor right at the soldiers that supposedly made up his army. He looked bored.

One of the men who was with Howell poked him, and called his attention to the man who led the party. Howell stared in amazement, and then turned to his mates. "That's the King! I am sure of it. Who else would dare to ride in front of the Commander-in-Chief?"

Small, round-faced Fredrick was bellowing, and waving around his sword as though he could somehow convince these sickly paupers that he was some great conqueror. This would-be warlord was almost upon them now. "Your King will stand with you as we defend the honor of Bohemia!"

Howell tried his Czech again, "Fight for Bohemia."

"What's Bohemia?"

"Cechy," Howell replied.

The soldiers stared at their translator. "He doesn't know the name of his own Kingdom?"

Apparently not. Howell watched as the mounted party continued along the column behind him. Fredrick, silhouetted by the setting Sun, swung his sword, and began to repeat his proclamation. "Soldiers! Your day of glory approaches!"

Howell looked right at the face of Sir Whore's Ass as he rode by. *'Count your days'*, Howell thought. *'You do not have many left.'*

# Chapter XXV

## The Battle of the White Mountain

Howell shook himself awake. Much as he had wanted to press on last night, even he had to admit that he could not go on without some rest. Giovanni had matched him pace for pace every hour of the way, and he could not overcome his need for some sleep, either. Howell's desire had to yield to necessity last night, and now that he was up, he found the Sun was much higher that he would have liked. Time was the one thing that he could not afford to waste. He had to move as quickly as may be.

He stood to examine what was around him. He was on the slope of a frozen ridge, watching the thousands of men of the Army of the Estates greet the clammy dawn. The steam from their breathing filled the air, and made the face of the ridge look like the flank of a great animal that had collapsed from exhaustion. The last two days would have been a punishing march for men in good condition, but these were a listless, skinny gang, desperately in need of a good meal. For many of these Bohemian rebels, this was their third Winter of fighting against the Emperor. For such as these, it was astonishing that any of them was still on his feet.

It was a revelation to Howell that so great a host could be gathered together in one place, for one purpose, under one man's command. He was awed by the sheer number of men within view. "I have never seen anything like this."

Giovanni looked at him groggily. "I wish I'd never seen anything like this."

Last night those members of Erlach's Company who had not dropped out from fatigue or hunger, had walked blindly up this slope before finally being allowed some rest.

"Looks like we're dead center in this formation."

Giovanni pursed his lips. "I'd like a big bowl of soup directly in my center."

*Um hm* Howell was completely absorbed by the spectacle around him. "Troubled as they are, this army will face the enemy today."

"Troubled, you say? This no-pay, no-drink, disorganized, starving, diseased, accursed excuse for an Army is certainly troubled. I thought I saw a man smile yesterday, but he was just gritting his teeth."

Giovanni was right. This Army was teetering on the brink, but there was still a spark left. "Look at the men around you, Giovanni. They have starved, and watched their friends die of disease. They have shivered throughout the Winter. They have just endured a march that the heroes of antiquity could not have stood."

"Heroes of what?"

Howell ignored Giovanni's question. "They did it because they supported one another. They did it because of the pride they felt in what they had become. They are bound together by that pride. Look at them. Look."

Giovanni did look. What he saw were exhausted men, but some of them were staggering down the slope to grab the arms of one of their comrades who had fallen out during the march, and had only now caught them up. He saw a man giving the last of his food to his friend lying on the ground. Men were checking their ammunition. Men were examining the state of their muskets. "I hope you're right," said Giovanni. "A lot of them look pretty spent."

They needed all the things that Giovanni had listed, and they needed the comradeship that Howell had pointed out, but above all they needed leadership that they could trust. Captain Erlach provided that. Unfortunately, that was as far as the men's confidence extended. They hated Sir Horace, the Commander-in-Chief that King Fredrick had placed over them, and thought little enough of the officers between him and their Captain.

Howell needed to do what was required of him, as well. He needed to resume his quest for Rosy. "Let us be on our way, mate."

Giovanni gave a great sigh, but, ever the good comrade, he fell in beside Howell.

Just behind this ridge upon which they stood were Prague and the Royal Court. Rosy was in that Court. Lovely Rosy. Loving Rosy. Brave Rosy. She was the best thing that had ever happened to him, and he wanted to be with her. He needed to stand by her side in this world of battle and cold and starvation and Sir Whore's Ass. His journey would lead him to a conclusion of his vendetta as well. Howell and Giovanni started off along the crest of the ridge. Just ahead Howell thought he saw a road that crossed over the ridge, and then led towards the Capital.

To one side, Howell saw the poor Army of the Estates preparing for the coming battle. On the other side were fields and burnt out villages and thousands of people. Some sheltered in the ruins, some huddled in makeshift tents, and some cowered in open fields.

The walls of Prague itself stood beyond this desolate land, but surprisingly close. Howell could make out columns of smoke from hearths in the city, perhaps making some food, perhaps just to provide some heat. There was an enormous castle rambling along the ridge line that was the center of the city. It was formed by an odd collection of mismatched buildings and towers and spires. In the middle of this palace Howell could see dilapidated scaffolding in front of an enormous church, with a prominent circular hole in its wall. This opening was filled with radiating stone mullions that still awaited their stained glass.

Then, right before him, beside the road for which he was aiming, he caught sight of a small party grouped around a small fire. They were the same men that Howell had seen riding along the column of marching men two days ago. At that time, they were trying to encourage the soldiers in a language they did not understand. Sir Horace. The King. Various aides. Boys holding horses.

Howell stopped.

Giovanni plowed into him from behind. "What was that, amico?"

Howell did not reply. He considered. He could not kill Sir Horace now for the same reason that he had not been able to kill him countless other times. The disorder it would cause might be fatal to the Army. It occurred to Howell, though, that the current situation could be turned to his advantage. He did not know where Rosy was, but this man almost certainly did. If he could keep Sir Whore's Ass in sight, the pig would lead him right to where he wanted to go. So Howell needed to be where he could keep an eye on Horace, but be as inconspicuously about it as possible. He needed to blend into all the soldiers around him. He would go back to Erlach's Company, and become invisible.

Giovanni had stepped up beside him and saw Sir Horace and the King as well. "Hullo. Look what the cat dragged in."

Howell turned, and put his arm on his friend's chest. "I think the men of Erlach's Company would like to see us again."

Giovanni knew exactly what was going through Howell's mind. He smiled. "They must be horribly lonely."

Howell found that he could keep this noble group in view even as he and Giovanni retraced their steps, and came once more to the men with whom they had made that damnable march. The men of Erlach's Company were standing. They had weapons in their hands, or most of them did leastways. They looked ghastly. They looked like an Army of ghosts risen from the dead. Howell wondered how many of them would give up even this shred of their lives before this day was done.

Howell looked at the campfire with Sir Horace and the King. What did they mean to do? Would they remain here throughout the battle? Horace should certainly stay, he was the Army's

Commander-in-Chief after all, but would the King stay or go? Did it matter? What mattered to Howell was Rosy, and for now that meant keeping this Army intact and Horace within view.

Erlach's Company stood on the right flank of their Regiment with less than two hundred musketeers. To their left were about the same number of pikemen with their enormous spears, and beyond them the other half of their shotte. Perhaps six hundred of the Regiment remained, probably fewer.

To Howell's right, on the other side of Erlach's Company, was a squadron of dragoons. They were mounted men who carried a sort of musket that was small enough to be fired on horseback, or, if they so chose, they could dismount, and use them to fight as infantry.

Beyond these cavalrymen, Howell could see that the whole of the rebel Army covered the face of the ridge. Each of the infantry Regiments was in a line of musketeers and spearmen like the Regiment that Erlach's Company was in. These regiments were arranged in a checkerboard formation across the front of the high ground. Laid out like this, each could protect the regiments on either side in both attack and defense. Some squadrons of cavalry were mixed in with this foote.

How many were there? Howell carefully examined the scruffy force that surrounded him. He counted fourteen regiments and twenty-five or thirty squadrons, depending on how one counted the Transylvanians up there on the crest of the ridge. Allowing for all those who had fallen out in the last two days, he estimated that the whole of the Army was about fifteen thousand men. For Howell, it was incredible to see the thousands upon thousands of these soldiers covering the hillside, knowing why they were here, and at the same time the prospect was grand beyond words. He was only twenty years old or thereabouts, and he was going to see a spectacle that would be remembered centuries from now. He stood in awed silence for quite some time, feeling the moment, absorbing it.

The sound of small arms fire now shot out of the dense fog bank at the foot of the slope. The Army came to life as thousands of men stood, and strained to get a view. There was nothing to be seen, however, and after a period of random skirmishing, several hundred Bohemian cavalry came straggling out of the fog, and disappeared to the Army's left. The Bohemians on the hillside fell back into their inactivity. One might even call it a daze.

Howell glanced at the gathering of exalted men just a bit farther along the ridge, and satisfied himself that Sir Horace was still where Howell expected him to be. Then Horace's group re-arranged themselves slightly. "It appears that Sir Horace is talking to the King."

"The King? Really?" Giovanni was much less interested in all this than Howell was.

Howell was not shy about looking at whomever he pleased. He shifted his focus from Sir Horace to Fredrick. This man had been Count of the Palatinate for ten years, but King of Bohemia for barely one. Most of the time that Fredrick had been King had been in the cold of one Winter or another, so he had become known as 'The Winter King.' Fredrick was bundled up in the same unmistakable bearskin cloak that he had been wearing when he rode along the column of his soldiers who were struggling to put one foot in front of another, struggling to save his capital, and struggling to understand his words in a language they did not speak. He was mostly standing idle. It appeared that His Majesty's official business was not all that pressing today, battle or no. The men in his group lounged around the fire, occasionally exchanging some lifeless conversation. Presently, King Fredrick mounted his horse, and wiped his sleeve across his nose. Sir Horace doffed his hat, and bid the King adieu. Fredrick began the short ride to Prague, followed by a small escort.

Howell watched them go. "It looks like his Majesty is off to have breakfast with the Queen."

"Maybe he'll bring us back a bun." Giovanni did not sound hopeful.

Now Howell observed the towering form of General Thurn riding across their front. When the Rebellion began, the whole of the Army had been in Thurn's charge. Now he had been demoted from commanding the Army to commanding the left flank. He was riding from that position to a rendezvous with the Commander-in-Chief, Sir Horace, at his headquarters next to the fire. These officers were joined by others of the Army's elite. It was not long before a vigorous discussion broke out.

"Giovanni, come have a look," Howell called. At this range it was impossible to understand the words, but the tone could have been made out by a deaf man. There was finger pointing and gestures of frustration, and Thurn slapped his hat on his thigh. After a quarter of an hour or so, the meeting broke up, and General Thurn rode past Howell to his post on the left with a scowl upon his face.

"That was encouraging." Giovanni used his most sarcastic tone.

Howell mirrored his dubious look. Other soldiers around them had seen the exchange as well, and muttered amongst themselves.

The fog at the foot of the slope slowly burnt off as the idle morning slipped away. What it revealed satisfied Howell's curiosity, but did nothing to increase his confidence in their leadership. Under cover of the fog, the enemy had positioned cavalry and guns and four massive blocks of infantry at the foot of the hill opposite Howell's comrades. As the concealment continued to evaporate, even more infantry appeared in a second line on both sides of the field. And then a third line on the left. There were swarms of cavalry, twelve guns, and all together, ten of the imposing infantry units. Each of these was a solid block of a thousand or more pikemen with their towering spears, and an equal number of musketeers in ranks and files on the front and sides that swelled the unit is dimensions. It seemed to be a pike formation with the musketeers deployed as nothing more than auxiliaries. Whether or not it was an effective formation however, the sheer size and mass of one of these units was terribly impressive. You looked at it, and instinctively felt an irresistible power.

Howell saw Captain Erlach walking here and there amongst his exhausted men, giving them encouragement. He thanked them for their sacrifice. He reminded them of the cause they were fighting for. He looked as tired as they did. He took a moment to look down the face of the ridge at the force his men would be facing, and then turned to resume his rounds through his Company.

Howell resumed his examination of the opposing Armies. It looked to him that there were about two thousand men in each of the enemy's huge regiments. That meant that the two Armies were more or less evenly matched in numbers. He was not so sure about a comparison of the fitness of the two forces, and he was very unsure about the quality of the leadership. The Emperor's forces were led by Count Tilly, a man with an excellent reputation earned in his many battles and his many victories. By comparison, Horace had no experience, and might even be in a treasonous plot with the enemy.

Giovanni fidgeted. "Maybe this is all just a show. Maybe they aren't going to attack us at all." He sounded far more hopeful than certain.

"Perhaps." Now it was Howell who did not sound convinced by his own statement. Howell hoped Giovanni was wrong. He wanted to feel again that awful thrill he had known when he had rushed the barricade at Pilsen, or faced down the cavalry at Zablat. He craved it with a fierce intensity.

*Boom!* One of the enemy's guns fired, startling the field with its heavy, echoing report. The speeding iron ball was aimed towards the chalk ridge that was being defended by the Army of the Estates. Howell stood spell-bound by his first experience on the receiving end of artillery fire. The ball hit the hillside a hundred paces in front of some Bohemian cavalry, and sent a shower of stone chips spraying out towards the horses and men. The ball ricocheted off the chalk, and bounded back into the air, traveling all the way over the cavalry, and landing again in front of the second line. It loosed another hail of shattered chalk, and leapt once more into the air. Howell lost track of its flight as it disappeared somewhere beyond the dragoons.

*Boom!* All of the enemy's dozen pieces of ordnance began to fire. *Boom, boom boom, boom, boom!*

Howell kept his eye on Sir Horace, but occasionally glanced back at what Captain Erlach's men were about. Some shuffled their feet. Some wiped their faces or their brows. Some talked with their mates. Whatever else they might be doing, they had steeled themselves, and were standing firm.

*Boom!* Imperial shot now flew everywhere amongst the Rebel soldiers. *Boom!* Every impact kicked up a geyser of lethal stone shards, and launched them slicing through the defenders. *Boom!* Horses reared, and men shied. *Boom!* The screams of the wounded rose across the hillside. The balls would rebound after every impact, and fly again about half the distance of their last bounce, and repeat the destruction.

It was terrifying. One had no defense, and no response. All you could do was stand, and watch as an iron shot bigger than your two fists clenched together came through the air to wreak whatever damage it chose. Those balls that were fired towards the flanks of their position made impacts that could just be seen, and not heard at all. Those that were fired towards the center, where Howell and Giovanni stood, could be seen for just the briefest instant before they arrived. Each shot seemed to be aimed directly at your nose. Then it tore up the hillside, and made an unholy noise.

Giovanni turned to look grimly at Howell. "You must be having one Hell of a time."

Howell could not really say that he was enjoying the spectacle, but for him it was terrible and compelling in a skin-crawling, awe-inspiring sort of way. The power that he was witnessing was unquestionably fascinating.

A cannon ball bounded over their heads with a deep hum, and both ducked instinctively.

"I hate this." Giovanni chewed his lip.

Howell could not help but smile.

Reloading the guns was a time-consuming business, so the stream of cannon balls was never really continuous, but its erratic nature was a trial in itself. When would the next one be fired? Where would it go? It became a species of grotesque entertainment for Howell.

After only a quarter of an hour, though, one by one, all of the Imperial cannon fell silent. There was a short span of blissful quiet, and then all twelve guns erupted together in a single roaring salvo. There was a shot in the air everywhere.

A ball struck four paces in front of Howell, and four paces to his left, directly in front of the Company. The *crash!* was terrific.

The air was filled with chalk dust and splinters of razor-sharp chalk.

Howell took a heavy impact on his chest and face, and was driven to the ground. His hat vanished. He instinctively covered his face with his hands. Hot liquid was running between his fingers. He shivered. He swiftly banished his confusion, and rationally considered what had just happened. He felt no serious pain, but he had suffered some goodly blows in his time, and had only felt the worst of it well after the event.

Men were screaming. Giovanni was standing over him, and making a tremendous amount of noise as well, and for some reason Howell wanted him to shut up. In order to get to his feet and silence him, Howell had to remove his hands from his face, so now he accidentally discovered that he had not been blinded. That was welcome news, to be sure.

"Vomitare l'anima!" Giovanni was gibbering, and jerking his arms. "I am going to puke up my soul!" He was staring at Howell in horror, and twitching his half-raised arms in uncoordinated frenzy. "Gesu Cristo! Howell! Amico!"

Howell looked down at his hands, and saw that they were covered in blood. He noticed a slash through one of the sleeves of his worn, grey coat, probably caused by flying chalk, and there was blood oozing from there as well. All down the front of his coat there was a long, bloody stripe. He looked like the very Devil, but felt only that he had been rapped on the head. What was going on? "Giovanni! Quiet!" He tried to settle his babbling comrade, but without success. Other men around the point of impact were yelling or staggering or writhing about on the thin turf.

Now he realized that he felt as if he had been punched in the chest as well as hit in the head. What had happened? He looked at the ground at his feet, and saw a musket. It was not his. There was blood all over it. He leaned over to pick it up, and the pain rushed to his head. He stood, banishing his dizziness, and enduring the headache. He examined the weapon. The cock was all bent out of shape, and blood covered most of its length. Then it all came together in Howell's mind. Someone in Erlach's Company had been hit by the cannon ball, and his musket had flown right into Howell. The muzzle had bounced off his head while the rest of it had smashed into his body, throwing him to the ground. It was sobering, even frightening, but not fatal. "Giovanni!" He struggled to get his friend's attention. "Amico!"

Giovanni tried to concentrate on what Howell was saying, but he was plainly overwhelmed.

"I am not hurt!" Howell yelled above the tumult about them.

"What?"

"The cannon ball did NOT hit me." Howell bellowed directly into Giovanni's face. "I am NOT hurt!"

"Not hurt?" Giovanni repeated the words without knowing what he was saying.

"No, friend, I was not hit. It was some poor bastard over there, and his musket flew into me."

Giovanni looked at the rest of the wounded men, and the light began to dawn. "So…" Giovanni pointed tentatively at Howell. "…all this is, is…"

"Yes, it is someone else."

"Oh Dio." Giovanni turned away with his hand over his mouth. "Gesu Cristo." If he had been fed, it would have all come up now, but as it was, he began a long siege of the dry heaves.

Howell had a roaring headache, and he would show a goodly-sized bruise presently, but he was alive. It was all because just an instant ago, a cannon ball had nearly killed him. A few paces difference in the line that shot had taken, and Howell would have been wounded, or even killed, like so many others had been. What did it mean? Death had come hurtling at him, and yet nothing had happened. Was he spared by the hand of God? Perhaps the Devil was keeping him for something worse. Perhaps he had been preserved by nothing more than the mindless path taken by a soulless eight-pound lump of iron.

Howell heard a commotion in the ranks. It was sweeping through the Company like fire in a field of ripe wheat. "Look!" The soldiers around him pointed, and yelled. "Shitte!" "Oh, God."

Howell looked out across the field. The whole of the enemy Army was advancing upon them. The simultaneous firing of all of the foe's guns had been the signal for a general advance. The enemy's big blocks of infantry crept towards them with the inevitability of an approaching thunder storm.

"Quiet in the ranks!" The men were shifting here and there, and muttering to one another. "Steady, lads." Erlach was doing what needed to be done to keep his men in order. "Take your places, and be silent." Erlach strode across the face of his Company, shushing those who could not stop whispering, and straightening out lines that were slow to form properly. Howell could feel the sense of panic subsiding. The men saw that they were not on their own. They were being given definite instructions from someone they trusted. They felt part of a unit again.

While the enemy's great regiments of foote continued their majestic advance, enemy cavalry threatened the Bohemian left. That was the area commanded by Count Thurn, and now he responded with cavalry of his own.

Howell was standing beside the Company, looking like a careless butcher, and watching the cavalry fight when Giovanni finally reappeared. Howell turned to greet him, and exposed his gory attire.

"Il mio Dio, Howell!" Giovanni exclaimed instinctively. "My God amico, you look, well, you look…"

Howell said nothing. He had composed himself, and was the determined warrior once again.

Giovanni tried to assume some of Howell's soldierly concentration.

Howell pointed down the slope and towards their left. "Our boys are doing rather well over there." A thousand or so cavalry were riding about, firing their pistols at one another, and swinging their swords. The foe's horse seemed to be giving ground. Many of the soldiers around Howell and Giovanni were watching the show anxiously.

Then a unit of Bohemian infantry advanced to support the horse. These were arrayed like the infantry around Howell, with all their musketeers on the left and right of the men carrying their great spears.

"That's Thurn's Regiment." One of the men in Erlach's Company recognized the flag they followed. Thurn's Regiment was the oldest and best unit in the Rebel army. Back in the days when Thurn had been the commander of the whole Army, the men of his Regiment had led in every fight. It looked as though Thurn intended to show Sir Horace who was the real leader on this battlefield. The Regiment's ancient flag was carried defiantly before the center of the formation. More than a thousand veteran infantry pressed forwards.

"Thurn will show them." Several of Erlach's men shouted encouragement. "Good old Thurn!" "Give them Hell!"

As though these shouts were a command, the front ranks of Thurn's musketeers began to fire. It was ragged at first, *pop pop popping* here and there. They were a good two hundred paces from the enemy, far, far out of range.

"What the Devil?" Howell and Giovanni exchanged worried looks.

The firing swelled. In a matter of a moment, the random popping was replaced with a sustained crackle from the front three or four ranks, and it did not stop there. The madness infected the whole of the Regiment. The musketeers in the ranks further back pointed their muskets blindly into the air, and wasted their shots. Now there was not a single loaded musket in the whole of Thurn's Regiment.

"Idioti! What are they doing?"

It was not just idiocy that had overcome Thurn's Regiment, but panic, as well. The useless volley seemed to have put the wind up the lot of them. Musketeers in the rear ranks began to run. Those in the front ranks tried to force their way through those further back who were tardy in losing their wits. Men were trampled to the ground. Muskets were discarded. Seeing their shotte abandoning them, the spearmen were now seized with the fever. As though a scythe had passed through the thicket of pikes, all those tall spear shafts now toppled over. A thousand of the best-regarded infantry in the whole of the army scampered off the field like so many sheep.

"God's blood."

"I think that they might have done somewhat better with the first act of our little drama," said Giovanni.

Howell and Giovanni were not alone in regarding this farce with concern, either. The whole Company buzzed. This was where Erlach made a tremendous difference, reassuring the men and preparing them to support one another. "Stand fast, men of

Erlach's Company!" Erlach raised his sword over his head to show that he was one with them. "Keep your ranks, men! Make me proud!" The men shuffled nervously, but they all kept their places.

A flurry of activity swept through the cavalry beside Howell. To his right, the leader of the dragoons called his horsemen to order. Howell watched as the cavalry were formed into ranks and files. They had to be as well aligned as infantry, but they had to do it while handicapped with clueless horses. Being a cavalryman was not an easy job. Then a bugle sounded, and the two squadrons began to walk down the hill. If lining up was difficult while stationary, keeping cavalry in formation while moving was nearly incredible.

Howell looked to the foot of the hill to try to work out the dragoons' objective. Tilly's advancing Imperial infantry line featured an odd angle. The two regiments on the left were noticeably farther forwards than those on the right. It seemed to Howell that the regiment closest to the center of their line was dangerously exposed on two sides.

The dragoons that had started beside Howell began to trot easily down the slope. They were aimed at the same enemy regiment that Howell had targeted.

The commander of that great block of foote saw the Bohemian cavalry approaching, and ordered his men to a halt. The regiment had to be re-aligned to defend not just the front, as was the norm, but one side as well. Howell was amazed to see them effortlessly shift to the new formation. Musketeers stopped, and turned to face to the side. Spearmen turned, and swung down their long pikes to hold them so that they projected well out in front of the musketeers where they formed an iron hedge to hold off cavalry.

Still, an infantry formation was most vulnerable in this sort of position, and the dragoons seized the moment. They bore down upon the enemy ranks, and opened fire with their short muskets while they were still in the saddle. The enemy musketeers fired back. Rank after rank, the dragoons rode up to the regiment, and discharged into the faces of the Imperial infantry. The foot soldiers recoiled, and the Bohemian cavalry kept pressing. The barrier of spears was incomplete, and the horsemen closed in, and began using their swords. Musketeers swung their muskets as clubs in an effort to defend themselves.

"Where's the support?" Howell said aloud. He looked at the idle infantry regiments of the Bohemian front line. Why were none of the Bohemian regiments of foote taking advantage of this opportunity? Now was the time to finish off that enemy regiment, while the dragoons had them back on their heels. "Advance, damn you! Advance!" Some of the soldiers around him muttered in agreement.

"Damn!" Howell focused on what was developing to the right of the struggle down below. "Look what's coming."

Enemy cavalry were streaming in to attack the Bohemian horse from the side and from behind. With the Bohemian dragoons engaged with the Imperial infantry regiment, they could not turn to meet this threat. Count Tilly had recognized the danger to his infantry regiment, and had responded, but the Bohemian commander had dispatched no one to further the dragoon's initiative. Field Marshal Sir fucking Whore's Ass was so unschooled in battle that he could not be roused from his lethargy to make an obvious move. Caught between the dogged defense of the Imperial foote and the slashing attack of the Imperial horse, the Bohemian regiment of dragoons could not hold. After only a few moments of struggle, the remains of the horse came thundering up the slope. The riders had their heads by their horses' necks for maximum speed. They were intent only on escape.

They were coming directly towards Erlach's Company.

Howell saw Captain Erlach standing impassively in front of the Company that was his duty, and his honor, to command. He yelled to make himself heard above all the racket made by the fighting. "It will be fine, lads. They will go around us."

Hundreds of fleeing horsemen were drawing nearer by the moment. Many were unarmed, hatless, or bloody. Scores of these terror-stricken refugees on their foam-mouthed horses came straight at Erlach's Company.

As the Captain had predicted, though, the men in the front ranks of the cavalry saw what was before them, and steered their horses so as to miss Erlach's men. As those cavalrymen turned aside, the horses behind them followed. They passed Erlach, but only barely.

They did not quite get around Howell. A horse tore past, and bounced Howell to one side. The stirrup fetched him a hard blow on his arm, nearly ripping off one of his sleeves, and hurtling him to the ground. The sound of pounding hooves was everywhere. He struggled to regain his feet.

Then it was over. The cavalry disappeared behind him. He stood for a moment, gathering himself. He looked down the length of the ridge. Sir Horace was gone! Howell had been so absorbed by the battle and by his own wounds, that he had missed when Horace had taken to his heels. Cowardly rat! Howell now had only one task. He immediately began to run as quickly as he could towards Prague and the Palace and Sir Horace and to Rosy.

Giovanni ran alongside Howell. They topped the ridge.

"I just saw the man of your vendetta disappear into that mob," said Giovanni.

Spread out before Howell was a roiling mass of refugees. Everyone thought that sanctuary lay within the walls. Howell knew better. With Horace's first plan in a shambles, his next move would be to kidnap the Royals, and to kill anyone who interfered with that. Rosy had never before faced such a challenge.

# Chapter XXVI

## Duty

Howell and Giovanni ran towards the castle, clawing and tearing their way through the struggling mass of desperate fugitives. These people had heard the terrifying sounds of the battle coming from the other side of the ridge, and fled towards the castle. This imposing structure sat atop a steep cliff. A road wound around the base of the cliff to end at the castle's gates. This road was choked with refugees frantically seeking entry to the castle. An even bigger mob was trying to get to the place where the road began.

It had taken the better part an hour for Howell to get near the head of this directionless, frantic herd. He had to win this race. He knew that he could not get to Rosy's side soon enough. She might fall prey to any of the madmen around him, but Sir Horace was the overwhelming threat.

The castle's great portal stood partially open, held out from the castle's wall by what must have been the combined effort of many men inside. Guards were blocking the narrow opening that remained, and only letting in a select few.

Howell needed to reach that gate, and get into that castle. The mass in front of him choked the road that he needed to use to reach his goal. He and Giovanni plunged into the maelstrom. They twisted, thrust, and heedlessly shoved people aside in Howell's rampage. Grunting and pushing, the friends fought their way towards the road that began at the foot of the cliff. Howell panted, his heart racing from his exertions.

They were at last nearing the road when Howell saw a rider forcing his way through the swarm. Howell knew instantly that it was Sir Whore's Ass that he was looking at. He was using the bulk of his horse and the point of his rapier to clear a path through the milling, shouting rabble. The army's Commander-in-Chief was working with all his strength and all his skill to desert his men. He reached the narrow road that led up towards the gate, and started up.

If Howell could get behind Horace, perhaps grab the horse's tail, he could follow in Horace's wake, through the multitude, and through the gate. But Horace was well out of reach, and getting further away with every citizen he knocked out of his way. He rammed some of the pack off the road, and these tumbled down the steep incline. Others he forced to flatten themselves against the cliff, and some of these were crushed. Sir Horace stabbed some with his rapier, and these fell from their wound. Horace's mount reared, and some unfortunate was trampled. Ahead of Sir Horace and behind him, the crowd continued to fight with one another for a chance to get into the castle. Their frantic yelling filled the air.

Sir Horace reached the gate, and passed the guards. They immediately stopped holding the gates open, and let the pressure of a thousand panic-stricken fugitives slam the doors closed. The commotion around Howell erupted to new heights as those in the crush realized that they had been denied entry. "No! No!"

Howell was filled with blind rage at seeing the object of his hatred escape. For a moment he cursed his repeated frustrations. Again! Again! Again! Then he stopped. He gathered himself. He focused. There was no time for distraction. He had to find Rosy, and it could not be too soon.

If he no longer could get in through the gate, he had to find another way into the castle. There wouldn't be any sort of gate or door at the foot of a castle. So, if he could not get through the castle's walls, he would have to get over them. Now his immediate objective was to find a way to the top of the cliff.

He and Giovanni battled their way back down the road through the roiling herd as they made their way along the base of the cliff. Howell had to find a way up. He had to. He passed sheer, featureless slabs that were impossibly smooth. Some actually had overhangs. He made the best time he could, pushing through the current of humanity as he moved along the base of the cliff, looking at every possibility that he found.

Then he saw a crack that looked like it had sufficient handholds that it could bring him to the top of the cliff. He could not take the time to examine it thoroughly. He didn't have the time to look for something better. It would have to serve.

"Amico!" Giovanni pointed to the crack. "Here. Climb."

Howell looked up at what he faced, and began to scramble up. He scraped his knuckles along the coarse rock. His sleeve caught and ripped. Giovanni cupped his hands at his waist to give Howell a rung on the improvised ladder he was building. Howell put his foot onto this step. He did not pause. He stepped on top of Giovanni's shoulder. Giovanni held his hand over his head, and Howell used that as well.

As soon as Howell started up, others around him saw what he was about. If this was another way into the safety of the castle, they wanted to take it as well. Dozens surged at Giovanni. He might have used his knife, but Howell could see that his friend just could not bring himself to descend to the level of Sir Horace, and stab his own people. Giovanni punched, and pushed to keep a space open around himself. The people in front of him might had stepped back at this, but they were pressed in upon him by the efforts of those behind. Howell could see that very soon it would not be just a matter of keeping this mob at bay, but a matter of survival. It was only Giovanni's tenacity and brawn that prevented him from being crushed against the rock behind him.

Using all the strength remaining to him, Giovanni forced a tiny space around himself. He used it to climb up the bodies of those before him. First, he stepped upon their knees, then he drove his toes into their bellies. The throng grew tighter. He was still in danger of being pinned against the cliff. He used his hands to push down on the heads of those who were closest him so that he could put his knees on the shoulders of them. He put one foot upon a shoulder. He pulled his other foot free, and put it another shoulder. Giovanni had his back to the cliff, and could barely keep himself upright in the press.

"Giovanni! Friend! Give me your hand," Howell called down through the storm-lashed sea of heads that extended out around him as far as he could see.

Giovanni could not turn. He reached up blindly. He swung his hand back and forth across the crack in the cliff that Howell had used to climb up. The mass of refugees scratched at him. Howell's hand felt Giovanni's arm, and he grabbed his wrist. Howell began to pull him up.

The mob below them gathered around the base of their route, and pulled down one another in their desperation to get to wherever it was that Howell and Giovanni were going. Howell was sure that Giovanni's arm must surely be wrenched from its socket. Someone grabbed Giovanni's ankle. Giovanni ground the back of this man's hand against the jagged rocks, and the man let go. Howell heard the commotion as the man fell back into the swarm.

Howell pulled while Giovanni scrabbled, and he finished the last yard laying on his belly. He took a moment, rolled over, and sat with his legs hanging over the lip of the cliff.

It was not just Giovanni who was cut by the rocks. The places where Howell could get a grip on his way up were very irregular. Some of them were very sharp. He cut the palm of his hand. He took the skin off some of his fingertips. By the time he had gotten Giovanni to the top, his hands were dripping blood. He wiped his bloody hand on his coat.

"Amico. Are you well?"

Howell hurt from a hundred blows and cuts. "I expect no worse than you." He focused. He could not delay. "We have to find a way in."

Giovanni had not really regained his breath yet, so Howell helped Giovanni to his feet, and they moved off at the best speed that Giovanni could manage.

They were on a flat lip of stone that the castle's wall was built upon. There was a generous path as they started, but soon it began to narrow. It narrowed further and further until they were forced to proceed with one of their hands upon the castle's wall. Then they came to a spot where there was no trail at all. The path that they had been following resumed on the far side, but there was a considerable gap where the castle's wall met the top of the cliff exactly. Here there was nothing but a sheer drop all the way to the plain below.

How could Howell continue? Perhaps there had been some sort of bridge here when they were building the wall, but there was no evidence of one now. Nor had they passed any planks that they could use to improvise one. Howell could see only one option. "Giovanni, back up a bit and give me some room."

"Mio dio, Howell. What are you thinking?"

"I must get to Rosy, and that means I must find an entrance." He would make this jump. He had to.

"But…"

"Back up I say. I am in haste."

Giovanni did as he was bid, and this gave Howell about ten feet to work with. "Are sure about this amico?"

Howell clasped Giovanni's hand. "Wish me luck, friend. You have been as true a comrade as any man could hope for. If I never see you again, know how much that means to me."

"It is an honor to know you, Howell." Giovanni wrapped his arms around Howell's shoulders. "Amico mio, you will succeed. I know it."

Howell turned to face the gap before him. He could only gain so much speed in the space in which he stood because the wall pressed in so closely. He would know in a moment if it was sufficient. He took a deep breath. He had eaten little, and run much. He had been nearly crushed to death by a mob of frenzied people who were as desperate to get into the castle as he was. Now he considered how he must accomplish this task with what little was left to him. It would have to be enough. He dashed the few steps available. He gained as much speed as he could. He pushed off. He flew. He took a risk. He put his foot on a tiny outcropping of rock below the defensive wall, and stayed in the air for an instant longer. This pushed him out from the wall, though. His belly landed on the far lip of the gap with his legs dangling in the air, and one of his arms dangling over the precipice. He pulled that arm in, but this threatened to push him back into the gap. He scratched at the ground ahead of him, and found a place where he could dig in his nails with one of his hands. He whipped his feet up and while his stomach was in the air, he clawed himself a bit further towards safety. Bouncing and crawling, he got himself fully onto the shelf on the far side of the gap. He lay still. He had be able to move on. Indescribable joy swept over him.

"Amico! Make way."

Howell turned over to see Giovanni standing at the same place from which Howell had begun his run.

"Giovanni. Maybe this is not a good idea."

"Anything you can do, I can do. Besides, you will catch me."

Howell smiled, and rolled over. He found a solid bit of rock on which to brace his foot. "Yes I will. You see that little nub of rock there? Step upon that as you go by."

"Very well."

Giovanni prepared himself. He was of a stockier build than Howell, so he wouldn't be able to get quite as close to the castle's wall as had Howell. He looked over the face of the defenses beside him to know where he could pass easily as he ran, and where he would have to be careful. "Ready?"

"Ready."

Giovanni launched himself. He stepped on the little rock projection that Howell had used, but he pushed off with more force. This meant that he landed farther out with only his chest on the far lip, with one hand in front of Howell and the other far off in space.

Howell grabbed his friend's wrist. Giovanni immediately began to slide back into the gap. He got his loose hand on the top of the lip to help pull himself up, but he was not making any progress. He began to scrabble his toes around, finding footholds. He could barely use them, though. He could not bend a knee to get a real push, so all he could do was push his toes down to gain an inch or two at a time.

Howell clung to Giovanni. He swung his foot so that Giovanni could grab his ankle, and get a firm hold. Howell wanted to lean over, seize Giovanni by his doublet and pull him up bodily, but Howell knew that easing off on Giovanni's arm would doom his friend.

Giovanni got high enough that Howell could see his eyes. Giovanni found another foothold, and shoved himself up a few more inches. As he did it, he snatched at a place further up Howell's leg. Now he had his elbows on the lip of rock. Now he scooted onto his belly. Now Howell pulled him completely onto the ledge. They lay side by side, panting, their lungs burning. They lay there, filling and emptying their lungs with the icy November air. The thin turf beneath their backs was freezing.

Howell sat up. "I must keep moving."

"Are you sure amico? It is so comfortable here."

Howell got to his feet, and pulled up Giovanni. "Let us be off." He began to walk along the base of the wall once more. It towered over him. He could not find a door, nor any place where he could climb. He had to find something. He had to get in.

He came around a corner of the wall, and found a confused tangle of boards and posts. What was this? Then he remembered seeing some scaffolding in front of the great church. He looked up at the huge rose window with its empty radiating spaces where the stained glass was meant to go. When the war had started, all work here had stopped.

The vertical posts of the work platforms were cocked at crazy angles, and the platforms upon them had been warped out of shape, all from constant exposure to the elements over many years.

"Giovanni. I must climb." Howell got his feet onto a board, and pulled himself up. "You wait here."

"You would leave me behind?" Giovanni said. "I am so disappointed."

The structure groaned as Howell started up. He looked for posts that were as near to vertical as he could find, hoping that these would stay up. Still, they wavered alarmingly. These posts were a yard or more from the church's wall to leave room for him to pass, and climb onto the next platform. Each transition was a danger, but he had to take the risk. He balanced himself on warped boards when he walked across to a different part of the scaffolding that seemed to offer a better route. He climbed, passing up from one platform to the next. Finally, he came to a place where the board over his head was at the level of the bottom of the window. One to go.

He started around the back of the rickety scaffolding to get onto this last board. He got a grip in the post just above the plank, and pulled himself up. Done! He looked through the window, and saw that there was a set of posts and boards on the inside too, just as he had hoped. He realized now how much he had feared that there would be no way down.

"Hey! Are you just going to stand there admiring the view?" Giovanni stood at the foot of the scaffolding, yelling up at him.

Howell got a good grip on the stonework of the great window, reached out, and grabbed one of the posts to steady it. "Use this one!"

Howell held on, and Giovanni quickly climbed. Soon they were sitting together between the stone mullions of the unfinished rose window. The scaffolding on the inside had not been exposed to the weather, so it was in much better condition than what he and Giovanni had just braved on the outside. As he climbed down, Howell found himself passing the largest cross that he had ever seen. When he reached the floor, he found the cross towered over a great altar. When he looked the other way, row after row of gothic arches stretched away from him. It looked to him like he was inside the ribcage of some enormous beast. At the far end was a great double door, bound with iron straps. This was the entrance meant to admit the worshipers on Sundays.

This would do for the unwashed masses, but the quality would have an entrance of their own. He looked to each side of the altar, and saw a door of regular size, and very well finished. He dashed over, and pulled this open. He found himself in a small chamber with the foot of a staircase before him. Giovanni piled in behind him. The Royals would have their apartments somewhere up there, and where Elizabeth was, there would be Rosy. Howell led his friend up these steps.

When they reached the top of the stairs though, Howell was faced with a new problem. Corridors led in both directions. He had no information, yet he had to act. Now. He knew he was going to have to take a terrible gamble. Which way should he go to find her?

"What's that?" said Giovanni. "Is that voices?"

Howell strained his ears, and then he heard what Giovanni had heard. It sounded like an argument. Who was it? What were they arguing about? He led Giovanni through several turnings in several corridors to follow the sound which got louder with every step they took. A woman was part of the argument. The language was heated and close to them. He came to an intersection of halls, and looked around the corner.

The Royals were gathered together in a sort of audience chamber with a great many columns. King Fredrick stood next to the foul Sir Whore's Ass. Queen Elizabeth stood near both of them. Sir Horace seemed to be the center of the discussion. Horace's thug hovered nearby, looking more like a guard than an observer.

Queen Elizabeth firmly addressed her husband. "Please Fredrick. Do not listen to him."

"Your Majesty." Sir Horace spoke in his best soothing tone. "Events have become, shall we say, rather uncertain. My recommendation is that you remain here until we know more."

Howell was certain that Rosy wanted this man dead in order to spare the life of her friend, Queen Elizabeth. Howell wanted him dead for his own reasons. The only weapon he had was the knife in his hand, which was little enough for an attack upon a man armed with a sword. It would have to do. He could improve the odds if he could get behind one of those columns.

The Queen tried once again to persuade her husband. "Fredrick, please…"

The King was exasperated at the badgering by his wife. "I will do as I think best, woman." He pointed with determination at Sir Horace. "This man is my councilor, my councilor of many years I might add, and I have followed his advice on many occasions." He said it with a note of finality. "I see no reason to change now."

"Of course husband, this choice is yours to make." Elizabeth did not sound convinced by her own words.

Just along the hall was a door standing partly open. This was a perfect halfway point as Howell moved to the place where he could make his attack.

Queen Elizabeth shifted her attention to Sir Horace. "Pray, Sir, help me to understand how the Army of the Estates, the Army that has been in your care these several years, has been so poorly supplied?"

Horace replied to the Queen's criticism, "Oh it is not as bad as all that…" and then he added as an afterthought, "…Your Majesty. At this very moment they stand upon an unassailable position, weapons in hand, ready to defend your Kingdom."

"Yet I hear a tumult just outside the castle's gates," said Elizabeth.

Horace was not to be deterred. "As I have urged repeatedly, if it pleases your Majesty to remain here, your safety is assured."

"Assured?" Elizabeth knew full well that this was an unvarnished lie.

"Hush, Elizabeth." Fredrick straightened himself, and stood a little taller. "It does in fact please me to wait upon events here. Myself and my household shall remain here as you have advised."

Howell waited until everyone was absorbed in their argument, and then he and Giovanni slipped along the wall, and through the door they were going to use as their halfway point. Howell plowed straight into a woman who was already there, and sent her sprawling. She was holding a baby, who immediately burst out in a piercing shriek. Howell saw who she was. "Rosy?" he said. He had found her!

"What in Hell was that?" yelled Sir Horace.

Rosy whispered to Howell, "I will run. Horace will follow. Kill him from behind." She sprinted out the door, the wailing child in her arms.

Just like that, their plan was made.

Sir Horace flashed by, hot after Rosy.

"I will deal with the guard," said Giovanni.

They both dashed out of the room and into the hall. Giovanni turned to begin his fight with Horace's thug.

Howell turned the other way to begin his chase, and found that Rosy and Sir Horace had disappeared. Where had she gone? It was his duty to find her. He bellowed out, "The big monument!" He meant the huge cross in the unfinished cathedral, but he could only hope that she thought the same thing.

He instantly started off trying to follow the crying of the child, but already he could hear that it was echoing, and making it difficult to be sure where it was coming from.

He would backtrack along the route that he had used to get here. Going back the way he had come might take longer, but he could be certain that it would get him the enormous cross. He steeled himself. H would be right. He had to be right. He began to retrace his steps. As he dashed off, he heard Giovanni yelling at the King, "Run man, run. Your army has broken, and he urges you to stay? It is madness. Get to a place of safety! Run!"

Howell ran down a corridor. He turned into a hallway, and then into another. Suddenly, he was at the top of the staircase that led down to the floor of the great church. The baby was wailing. He dashed down and froze. The back of the enormous cross and altar were lit by the lowering Sun. Their deep, sharp-edged shadows were centered in a pool of light on the floor surrounded by the darkness of the rest of the grand church. The child was crying somewhere directly before him. He rubbed the sweat off his forehead with the back of his hand, leaving a bloody streak. He heard the scrabbling of feet on the flagstones that made up the floor.

*Ha, ha*, "my dear. You may elude me for a time, but you know full well that in the end I will run you down, and put my blade straight through your heart."

"Oh, Sir," wailed Rosy. "Show pity. I am but a poor serving wench, forced to do the bidding of others." She was stalling. She hoped that she was at the 'big monument' that Howell had meant. Sir Whore's Ass was backing her into a corner. Time was running out.

Howell began to move from one pillar to another, being sure to move only when Sir Horace had his back to him. He was duty-bound to serve Rosy, and relishing the thought of killing the foul bastard he had pursed all this time.

Rosy had her eyes locked upon Sir Horace with the same penetrating intensity that Howell had admired when they saw off the pirates together. Her auburn hair was brilliant in the Sun, while all around her was darkness.

Howell was faced with the same problem that he had faced just a moment before. Horace had a sword, and Howell had only a knife. What could he do to win this fight? Howell had to put an end to this. One false step, and Sir Whore's Ass would be upon Rosy. He would drive his rapier through her. Over and over. He would strike repeatedly until he was sure that Howell's love was dead.

Suddenly, Howell knew what he must do.

He let out a blood-freezing, yodeling wail that echoed throughout the cathedral like the screaming of those condemned to eternal torture in Hades. The child joined in, louder than ever.

Sir Horace whipped around, searching for the source of this awful sound.

Howell waited just a moment. When Horace was looking in the wrong direction, Howell sidestepped quickly into the light. One side of him was utterly black, and the other was blinding white. He remained utterly still. When Sir Horace again swept his eyes over this part of the church, it looked as though Howell had appeared out of thin air.

Whore's Ass regarded Howell in bewilderment. "I… I… I saw you die." The man he had paid to have killed now stood only a few paces away from him.

Howell raised his stiffened arm ever so slowly to point his finger at Horace. His knuckles dripped blood from his scrabble up the rock wall to get to the foot of the castle. He held it there for an instant, and then bellowed, "Murderer!"

Rosy looked for just an instant, and then her eyes wandered crazily as though she was trying to work out something. She looked at Howell again. Then she came to understand what he was doing. She burst out in an ear-splitting, terrified scream. She pointed at Howell in horror as she got another lungful of air and shrieked, "A ghost! God save us! A ghost!"

Sir Horace gasped. His jaw dropped, and his eyes opened wide.

The parts of Howell's face that weren't covered in dried blood were streaked with fresh gore. He had been splattered with other men's blood, and doused in chalk dust during the bombardment.

His clothing was ripped and stained, and must look like the apparel of a half-rotted corpse.

Sir Horace had 'seen' Howell stabbed to death, so Horace accepted the evidence of his eyes, and believed that Howell had returned from the dead to haunt him.

As soon as Rosy saw that Howell's act as the vengeful spirit was holding Sir Horace's full attention, she silenced her little bundle by putting him to her breast. She crept down the space between the pews she now occupied, all the while careful not to do anything that might attract the notice of the sword-armed man in the room. She stepped behind a column, and disappeared.

Howell widened his eyes, assumed a stern expression, and whispered, "Thou hast sinned."

Horace's jaw began to quiver, and he took a step back.

"Thou hast done me murder."

Sir Horace's hands trembled, and the rapier shook in his loose grasp. "No, no, it was not me."

"Liar!"

Sir Horace did a little hop in place at Howell's shout. He did not drop the sword, unfortunately.

"Judas accepted thirty pieces of silver, and thou hast offered the same to those who wouldst do thy killing for thee." Very menacing and all, but where was he going with this haunting that he had just invented?

Horace took another step back.

Howell matched him with a step forward, which put most of body into the brilliant light beside the cross's shadow, but he positioned himself so that his face fell completely into the darkness. His hand was covered with chalk dust, so he kept it out where it was brightly lit. It only served to better hide Howell's face, and to enhance the other-worldly atmosphere. Howell kept his index finger pointed directly at Horace's face.

Sir Horace's sword twitched in his hand as he fearfully backed away from the 'ghost' before him.

Howell matched Sir Horace's slow movements in eerie synchronization. He did not want Horace to get a good look at him though, so he followed slowly to keep his face entirely in shadow. Only his bright, menacing fist was fully illuminated. "Forces from beyond the grave have sent me to offer you a chance at redemption."

"A chance?"

"A *final* chance to prove yourself worthy of salvation, murderer."

Sir Horace bumped into a column.

When he stopped, Howell stopped at the same instant, his ashen hand blinding in the Sunshine. His fist seemed to glow, which made it even more difficult for Horace to see the rest of him in the shadows. Blood dripped from the cuts on Howell's fingers and palm.

"Spirit," stammered Sir Horace, "who has sent thee?"

Perhaps the man smelt a rat. Perhaps he was only trying to discover with whom he was bargaining. In any event, Howell's performance as the vengeful spirit was beginning to fray around the edges. Who was he going represent? A saint? What Saint's name could he use? Justin! St. Justin, of course. Giovanni had said that he was under the patronage of St. Justin, and he had tracked this monster with this always in the back of his mind. "Will you do the penance offered by St. Justin?"

Sir Horace's face lit up with joy. "St. Justin, forgive me!"

"Then prepare thyself. To accept divine protection, you must cast off worldly defenses."

Sir Horace hesitated before realizing that he still held a sword. He dropped the blade as if it was red hot.

"Under the protection of St. Justin, you will no longer require such implements of death. Kick away your useless tool."

The sword rattled off into the darkness. Howell did not want Sir Whore's Ass to have a chance to get his rapier again as the rest of his plan unfolded. He stepped out into the light so that he was fully visible. He pulled his knife, and tossed it behind the altar.

Sir Horace looked at him in confusion. "You are the ghost of the man I saw killed, are you not? Defoe? Donner? Dickens?"

Howell dropped his other-worldly tone, and said conversationally, "Davidson. You really should have learnt my name, Sir Whore's Ass, and now you will fight me man to man." He stepped forwards, and seized the front of Horace's doublet.

"You murdering shitte. St. Justin may forgive you, but Howell Davidson will not."

Suddenly the truth thrust itself into the gentleman's mind, and he attempted to step back, but Howell kept his grip on the man's jacket, and swung his fist hard into the side of Horace's head. His opponent fell onto the flagstones.

"Get up, you turd. Get up. You would murder my woman? You would send me to the gallows? You would shoot me? You would watch as I was murdered? I will have my revenge, you cockroach, and you will supply it." Howell would be repaid for all of this. He poured all of his hatred of this man, and all his pent-up fury and frustration into this fight.

Sir Horace was suddenly on his feet, and just as swiftly pulled his dagger from its scabbard at the small of his back. He drove it straight towards Howell's stomach.

Howell leapt to avoid this unexpected thrust, but the blade's tip scraped across his belly just the same. He had entirely forgotten Horace's dagger. He leapt again, to keep just outside the reach of the blade. Horace faked a thrust, which Howell evaded. Horace watched Howell's eyes, and Howell kept his eyes on the dagger.

Horace thrust again, and missed again. They circled one another. Horace kept his eyes on Howell, but yelled towards the altar, "You red-haired slut, you will watch while I kill this maggot that you love, and then I will kill you. You will pay for all your meddling." He moved from side to side, forcing Howell back towards the steps leading to the altar. "Ghost, eh? You will soon be a real ghost." He made another quick jab.

This time Howell used it as an opportunity. He slapped the blade aside, and sprang forwards. He used both of his hands to grab Horace's wrist. Howell held the blade away from himself as his foeman struggled to regain control. Sir Horace had a free hand, and he used this to punch Howell in the head. Howell saw the blow coming, so he was able to cock his head, and prevent a solid blow, but his ear suffered just the same.

Howell raised his knee and smashed Sir Horace's arm down upon it. Then he did it again. Sir Horace yelped. Howell did it once more, and the dagger fell to the floor. Should he let go of Sir

Horace, and pick up the blade? He never had the chance. Horace reached out with one toe, and kicked it away.

Now Sir Horace twisted them both about, and drove Howell's back forcefully into one of the cathedral's columns. It knocked the wind out of him, and loosened Howell's grip on Horace's wrist.

Horace pulled his hand free. He dug his fingers into either side of Howell's neck, and began to force his thumbs into Howell's wind pipe.

Howell drove one of his fists into the man's belly. Then he drove in his other fist.

Sir Horace let go of Howell's neck.

Howell kept pounding the man's stomach until there was enough room for him to start hammering his head. Back and back they moved, Howell landing blow after blow. They came to the first step around the altar, and Sir Horace sprawled upon the full array of steps.

Howell was upon him in the blink of an eye. His hands were big enough that he was able to clutch the whole of Horace's neck.

Sir Horace's hands shot up to his throat. He tried frantically to break Howell's grip.

Howell was nose to nose with the man he hated most. He looked down at the face of the man who had done his best to murder him over and over. Sir Whore's Ass had tried to get him hanged so that he could watch Howell strangle. Howell would make the same gift.

Sir Horace's eyes bulged. Howell was not to be denied. Time and breath grew short. Howell squeezed as tightly as he could. Horace's hands fell limp onto his chest. Howell grit his teeth, and clutched Horace's neck with terrible fury. Horace's eye lids drooped, and he ceased making the slightest movement. Howell shook the man's head like that of a child's rag doll. Howell held onto the man's neck so long as strength remained, but eventually his arms began to quiver with fatigue, and he was forced to let go. He rocked back onto his heels. "And you never bothered to learn my name. It is Davidson, damn you!" Howell slapped the side of the corpse's head. "Davidson! Howell Davidson!" He turned from straddling Sir Whore's Ass, and rolled over to sit upon the step next to the man he had sworn to kill. Howell gave a great sigh of relief. He had gotten his revenge. At last. At long last. Was it really over? Had he finally gained the vengeance of which he had dreamt all these years? Had he satisfied his vendetta? Yes, it was true. He savored the reality.

Rosy came to him. She looked down with evident satisfaction at the man that Howell had just executed. Only a moment before, he had seen her lock her eyes upon her assailant in complete defiance. Then she had cleverly distracted Horace's attention just when Howell needed it. The curves of her hips beckoned. The front of her dress hung open; the unmistakable swell of her breasts visible in the gap. At this moment, Rosy was everything that Howell had ever imagined a woman could be.

The child she carried on her hip made a happy noise.

Rosy raised her head, and looked Howell squarely in the eye. "The jewels."

"Jewels?" Howell's train of thought was thoroughly scrambled.

Rosy scurried around the altar to look at the back. "Those dolts! Had no one the wit to collect this?" She started tugging at something. With the baby occupying one arm, she was having some difficulty doing whatever it was that she was doing. Howell got up, and walked around the altar to help. There was a panel on the back of the altar that would have been difficult to see unless someone had shown it to you. She was trying to get it open.

He crouched down, and retrieved his knife. "Here. Let me help." He tried to discover how to move the panel.

Rosy looked around to see if anyone had come into the church. "We must get the treasury to safety."

"Treasury?" Howell looked up at Rosy. "This is the treasury?"

"What better place for the Crown Jewels than under the protection of God?"

"Crown… Crown Jewels?" Fabulous treasure? Limitless wealth? Howell tugged at the panel, his brain suddenly clogged with visions of cavorting in feather beds, rivers of beer, fine horses, lavish raiment, roasting meat, and a palace that was the envy of the Sultan of the Turks.

The panel in the altar came free, and Howell pulled it full open. He gazed into the chamber. The whole of the altar was hollow. There were chests upon the floor and shelves jammed with boxes and bags. Some of them were open, revealing necklaces and rings and earrings.

Rosy instantly stepped away, and just as quickly returned with a large piece of cloth. "Here!" She threw it to Howell. "Put it all in here. Take it! Take it!"

Howell did not require any more detailed instructions. Furiously, he began making a heap of glittering wonders in the middle of the fabric. Bags bulging with gold coins. A crown. His head began to swim, perhaps from repeatedly looking up and down, or perhaps from dreams of a lifetime of sumptuous luxury. Broaches. Pins. A ball with a gold cross sticking out of the top of it. Sticks with gold knobs on both ends. Adornments of every kind. Baubles that he could not even put a name to. Strings of white beads.

Rosy had been doing what she could to help with her free hand, and now they bumped into one another as they both worked frantically. They grinned at one another in pure bliss.

Even their little companion seemed to be enjoying the moment. *Gumaflibavif*

"You are absolutely filthy!" laughed Rosy.

"Filthy rich, you mean." Howell was holding in his hand two magnificent necklaces, each filled to bursting with dazzling stones of captivating, beautiful colors. He jammed them through the gap in Rosy's bodice, one on each breast.

*Ooo!* She squealed.

*Oegaga!* agreed the child.

Howell laughed joyously, and Rosy latched onto him with her free arm to crush her lips against his in a kiss of such intensity that Howell momentarily forgot that the treasure in his arms could wait, while the treasure around him could not. The child was threatened with suffocation. Howell could not have been happier.

Then Rosy released him, and Howell returned to his work.

"Do not forget the pearls," she said, pointing at the white beads.

He tossed in the pearls. He gathered in box after bag. He emptied shelf after shelf. The heap was bigger than his arse.

"All right, Howell, for pity's sake, let us go!"

She was right, of course. Someone might appear at any moment, and they certainly did not want to share this loot. He gathered together the corners of the cloth, tied them into knots, and threw the bundle over his shoulder. Things fell from the sack, and rattled onto the floor behind him, but he did not bother to investigate. Rosy led him through the door that she had used to enter the church. When they entered the corridor though, she turned in the way opposite to that which she had used when she was being chased by Sir Horace. Sir Dead as a Cobble Stone Whore's Ass. The hall made several turnings, and they passed doors and intersecting hallways, but Rosy never slackened her pace. They went down a staircase.

At the end of a short passageway was a door to a courtyard. Waiting in this enclosure were horses and carriages and people and King Fredrick himself, standing in his bear-skin cloak, and looking about in profound confusion. It seemed that Giovanni had succeeded in his mission and convinced the dim Fredrick that escape was his only option. In the outer wall, Howell could see that there was a second, smaller exit that had not been discovered by the mob that was tearing itself to shreds outside the main gate.

Rosy started out the door that led to the courtyard.

"Hold!" Howell skidded to a stop, and grabbed Rosy by the shoulder to stop her, too. "That's the last place we want to go. This way!" He tried to drag her away from the door.

Rosy, however, was rooted solid, and Howell lost his balance as his fingers lost their grip. Rosy stood in the doorway, holding the baby. The light streamed in around her, an exquisite silhouette. Her red hair seemed almost aflame. "Darling, I have to make sure Rupert is safe."

The little rascal beamed an innocent smile at Howell. *Foosesamuh*

"Of course." How long could it take? Drop the baby with one of the maids, and they would be gone. Then he and Rosy and a sack full of treasure would be off to a future of wealth and luxury and pleasure. *Ah!* He unobtrusively set his load of booty on the floor behind him. The best place of all to deposit the child had just arrived.

Bustling down the hall towards them was a small party, and leading them was Elizabeth herself, Countess of the Palatinate and ever so briefly Queen of Bohemia. She was as beautiful as he remembered her from the night of the masked ball in Pilsen when she had pretended that no one would recognize her if she called herself 'Miss Albion'. Even in the haste of fleeing a disaster, she radiated elegance and charm. Her air was that of someone about to start on a grand adventure. She looked at Rosy. Her eyes grew wide. "Oh, my dear. I'd completely forgotten about dear little Rupert. Thank God you are so attentive."

Rosy nodded, and gave a miniature curtsy. "You saw that I had to flee from Sir Horace. He was threatening both of us with murder when Mr. Davidson arrived, and assured the survival of both Prince Rupert and myself.

"Survival? Oh, my goodness," said the Queen. She was genuinely concerned.

*Mfugahshig*

Howell observed the Queen with the eye of a practiced lecher, and decided that Rosy had the better figure.

"And Mr. Davidson, here…" Rosy indicated Howell. "…has rescued the Crown Jewels."

Howell was in the midst of bowing to the Queen when he froze, halfway between disbelief and having just been kicked in the balls. Did she really say that? What was she up to?

"Your Majesty, may I present Mr. Howell Davidson. You may remember him from our walk to the masked ball in Pilsen." Rosy placed her free hand on Howell's grimy, tattered arm.

"Any friend of yours is a friend of mine, dear," said Elizabeth, repeating the excessively bold reply that Howell had used on the night of their masquerade in Pilsen.

Rosy smiled proudly at Howell before speaking once again to the Queen. "Through bravery and quick-thinking he has rescued the Crown Jewels of Bohemia, your Majesty. Your jewels, Elizabeth."

Howell stood there, a ragged, filthy, bloodstained scarecrow with one hand holding the neck of a half-concealed sack. One would be forgiven if they thought that he was a thief. He attempted to finish his bow. His mind was spinning.

Elizabeth looked at the gentleman in her little company. She indicated Howell's baggage.

Rosy tugged on his sleeve, and Howell realized that she wanted him to bring forth the cloth filled to bursting with the fruits of their ransacking. There must be a cunning plan unfolding here, but for the life of him, he could not see it. Why didn't Rosy hand over this Prince Rupert to his mother? The heavy sack slid from behind Howell's leg, and came to rest before him. As it bulged, its cargo of valuables could be glimpsed through the gaps. There were the tantalizing flashes from jewels and the sunny glow of gold.

"Well done, lad." The gentleman shook Howell's hand for one heart-felt moment, and then grabbed the bag. A metallic rustling sound came from the pack as jewelry shifted position and gewgaws clattered into goblets. The man hurried off towards the coaches with the valuables.

The stupefied look on Howell's face exactly mirrored the confusion inside his head, but the Queen mistook it for awe in the presence of Royalty. Elizabeth beamed as she stepped up to Howell, and placed her hand upon his. "Mr. Davidson." She said it with great warmth, as though she had been anticipating this meeting for a long time. "In all this confusion, only you have kept your head. Only you have held fast to honor when all about you have been consumed by fear and by greed. My knight."

This is all part of a ruse, Howell told himself. Rosy will drop the other shoe in just a moment. He glanced out the door but the swag had already disappeared.

"I do not know how the Royal Family of Bohemia can ever thank you sufficiently, Mr. Davidson." The Queen removed a silver ring from her hand, and offered it to him. "As a token of our thanks, please accept this."

Howell bowed more deeply this time. He extended his grubby paw, and the Queen deposited the ring in his palm. As he straightened up, she inclined forwards, gracefully locating the cleanest part of his filthy face, and lightly placed her lips upon his cheek. "My friend." Her eyes gazed into his for a last sorrowful moment, and then the Queen turned away. She stepped out into the daylight, and began her long journey into exile.

Howell and Rosy were alone in the corridor, not even a hastily-assembled mass of plunder to keep them company. It was really gone, then. The honest wench had not been executing some diabolical scheme after all, but then, playing fair with one's friends was an impulse that Howell heartily endorsed. Starting their new life by stealing from her longtime companion was probably not the best course. Rosy's lips were starting to pucker, and her eyes were starting to tear. She gave a loud sniffle.

This poor-but-honest pose was uncomfortable for Howell, too, so he attempted to make light of it and managed a sheepish grin. "It was just a little kiss."

"Oh, Howell," she cried, and threw herself into his arms. "Howell, I do not care about that! It is nothing to be kissed by Elizabeth."

Really? As he hugged his sobbing woman, it seemed to him that getting kissed by a Queen, and a beautiful Queen into the bargain, was the sort of thing that a man could brag about until the day he died.

Rosy sobbed as she pulled herself back so that she could fix him with her tear-filled eyes. "I will never forget you, Howell."

"Forget? What do you mean forget? Just give the baby to his mother and…"

Rosy took a step back, shocked at what she had just heard. "No, Howell, you do not understand."

"No, darlin', *you* do not understand. You are handing off junior, here, to his ever lovin' mum as a sort of free gift with crown jewels, and we're getting the Hell out of here."

Rosy stood before him, Prince Rupert in her arms. Tears were sheeting down her freckled face. Just as Howell was about start in on her again, she blurted out, "You must swear never to repeat what I am about to tell you."

"What? Swear?"

"Just do it!"

Howell was taken aback by this emotional creature before him. One moment a sobbing blob, and the next a she-wolf with teeth like nails. It seemed best to just go along with whatever it was that she needed so badly. "Yes, of course, Rosy. I won't tell anyone."

"Do you swear?!"

"Yes, Rosy, I swear."

Now the anger was switched off, and she was all tears again. Amazing. The child was quiet for once, but he was watching all this with the greatest interest.

"I am not Rupert's nurse, Howell." She swallowed hard. "I am his mother."

"His… What?"

"And you are his father."

For perhaps the only time in his self-assured life, Howell Davidson was struck dumb. His mouth was hanging open, true enough, but no sound was coming out. He looked down at the boy in Rosy's arms.

The lad looked back. Rupert's tongue was out and he was grinning like he did not have a care in the world. *Rhowicob*

"When I came to visit Pilsen last April…" Rosy's face began to brighten with the poignant memory. "…and you were so very clever to find me, and we spent that wonderful night at Elizabeth's 'secret' reception. Do you remember?"

That night she had knelt naked before him. She had promised to do anything for him if he wouldn't try to kill Sir Horace at the masked ball. Anything. She had kept her word. "Yes, of course I remember."

"Well Elizabeth, I mean Her Majesty, the Queen, became, well…"

Howell just stared at her, trying to make sense of her rambling tale.

"She became with child soon after, *um*, I mean when I, *uh*, became so. Our son was born, and then two days later came Rupert. The real Rupert, not our Rupert," and she held up the baby so Howell wouldn't be confused. "Anyway, the real Rupert was still born."

Her rush of words stopped at that point. Howell could read in her face the sadness that had filled the Palace when the King and Queen found that their third son would not get even a single day upon this Earth.

"Sir Horace decided that it wouldn't do to give the people bad news. It was just after the coronations, you see, and they were trying to make the people feel good about…" She trailed off and looked up at Howell.

He waited patiently for her to finish whatever it was that she needed to say about his child.

"So Sir Horace wanted to pretend that our son was their son. And I said yes."

Howell had never been sure about how he should feel about Rosy. She had been adventure and joy, to be certain, and he had felt something much deeper that he had supposed must be love, but now there could be no doubt. He reached out, and gently encompassed his woman and his child in his arms. It was as though he had been tested with the crown jewels, and was now being rewarded for doing the right thing. He just held them in great contentment.

Rosy's composure was entirely gone, and she sobbed disconsolately. "I did not know if I would ever see you again! My son, our son, could live as a Royal Prince. He had never want for anything." She ripped herself away from Howell, and said defiantly, "What kind of life could I have given him, compared to that?"

"You would have given him a mother's love, just as you are now."

"And now it is all gone to pieces!" The wonder of it was that she still had tears to shed. "I made this terrible decision so that he could be happy and safe and now this, this war!"

"What's done, is done, Rosy." Howell reached out his arms once again. "If it is exile for us, then so be it."

"No!" Rosy shrieked like he had stuck her with a dagger. She flattened herself against the stones of the wall, and regarded him as though he was a monster. "You cannot come with us! You are not even supposed to know." She took one last, hard look at her man through her puffy eyes, and bolted for the door.

Howell caught her, and spun her around. He held her by her shoulders, and spoke to her averted face. "Go now in peace, Rosy. Know that I love you, and that I love our son." She plunged her head onto his chest.

"We will meet again, Rosy. I know it. I promise it."

She held him for one last moment, and then fled. Rosy scampered down the stairs, and Howell followed as far as the doorway. His woman and his child hurried across the cobblestones towards the waiting carriage. Rupert, his Rupert, regarded him over his mother's shoulder. His lips were moving, but Howell could not hear what nonsense he was babbling this time. Rosy's striking red hair moved farther and farther away. The head that bore that hair never turned to look back. She climbed up the carriage's steps, and disappeared from view. In a moment, the driver whipped up the horses, and Howell's family was gone.

Howell was at a total loss. None of this made the slightest sense. A half an hour ago, he had vented his blood lust on a man that he had hated, and pursued, for what seemed like his entire life. A quarter of an hour ago he had a fortune hanging off his shoulder and a delectable, auburn-haired seductress to help him squander it. In half that time he had given away the treasure but had discovered that he had a son who was as ready to state his opinion as his father. Now he was penniless and alone. It was crazy.

No, it was not crazy, it was infuriating! He had done everything that he was supposed to do. He had concluded his vendetta, which eliminated a man who was a curse upon mankind. He had buried his hands in a fabulous treasure, and then he had done the honorable thing. He donated his ill-gotten gains to Elizabeth the Fair, and received the Queen's gracious thanks. He had done everything for Rosy that could be expected of a man, and she had shown him that he had a son. Not just a son. A prince. And then she had left him.

He'd done it all and for what? Howell surveyed the now empty courtyard. How could he have been so stupid? She told him to give away a King's ransom, literally enough money to buy a King, and he had done it! She had told him that she could not stay with him, and he had just kissed her goodbye. He'd had everything, *everything*, right there, in his hands, and he had just let it all go.

Lost in trying to make some sense of what had just happened him, Howell did not see a lone figure striding across the courtyard that had just been used by the escape of the Royal Court. This person would have been familiar to Howell if he had looked, but he was blinded by visions of what might have been, and drowning in regret and anger. He was therefore taken by surprise when Giovanni jammed his fist savagely into his arm.

"Deserter! You told me to get the Re and Regina on the road, and then you just cast me adrift to fend for myself."

"Giovanni!"

"Si, it is Giovanni. Made his way here without any help from you!" He pushed Howell's shoulder which made him stagger back a step. He took a swig from a bottle of fine wine that he had pinched from the dining room table, and made to sit down on the top most step. "So, did you find her?"

Before Giovanni could take a seat, however, Howell jammed his face into that of his friend. He transfixed his mate with a fierce stare. "Yes, damn it! I found her. She was about to get raped or killed or both by Sir bleeding Whore's Ass, and I stepped in!" The note in his voice was not that of triumph.

*Err* "That sounds pretty good."

"Good? Good?! It was brilliant!" Howell's face was contorted by fury. "I pretended to be a ghost, and convinced the bastard to drop his sword, then I beat him bloody, and squeezed the life out of him!" Howell shook his fists under Giovanni's nose, imitating the execution. "And then Rosy shows me the treasury, and I fill up a cloth with the Crown Jewels!"

"Oh, please! Pull the other one."

Howell stalked out into the courtyard to get a breath, and turned back to lecture Giovanni further. "She has me haul down a cloth full of necklaces and gold coins and rings and then she tells me that Rupert…" He caught himself just in time. He may have been thoroughly buggered by honor and duty and being a good little boy, but he had sworn, and he could not break faith with Rosy.

Giovanni was listening to Howell's saga, but seemed to be thinking about something else, too. His eyes searched the courtyard, and peeked into the door of the Palace.

"…and then Rosy tells me that serving the Royal family, and doing for the Prince is more important that coming with me. Rosy's gone! She was duty-bound to protect little Rupert and flee into bloody exile with the bloody Royal family, and she just left! 'Ta, ta, Howell! Godspeed, Mister Davidson! Thanks for all the jolly times, and fare thee well!'"

"Real crown jewels?" Giovanni had a look of keen anticipation on his face. "So where…"

"What? The treasure? The sack jammed to overflowing with jewelry and coin and little crown thingamajigies?"

"Esattamente!"

"I *gave* it to them!" Howell exploded. "My idiot woman says, 'Oh Howell, my hero! Give the jewels to her' and I just *gave* them to her."

"Her being…"

"The Queen! 'Here, your bloody majesty. What do I need with a bag of gold? You take it. Please! Off you go, now.'" Howell looked around the empty courtyard, his fury marking his face. "I had it and I *gave* it away. I had Rosy, and she left me."

Giovanni walked up to stand next to his amico. He handed the bottle to him.

Howell took a long pull from the jug. "Rosy and the Queen both kissed me goodbye. Blutworth gambled with my life at Zablat. Horace and Vlok and our Captain and Mansfeld all tried to kill me." Howell took another swig from the bottle, and looked at Giovanni. "In all the mess of my life, you've been the one person that I could count upon." He had been by Howell's side in every scrape they had shared and now, for the second time, he had done what was necessary to maintain their team. "You've been a capital mate." Howell extended his hand. "I told you how much I appreciate your friendship an hour ago, and I am happy to do it again now."

Giovanni grinned, and took Howell's hand. "That's just meraviglioso amico, but I'd appreciate it more if you would tell me something altogether different."

"What?"

"That altar full of treasure. Did you and the girl clean it out completely? You know, every little nook and cranny?" *Eh? Eh?*

Howell's face lit up like a bonfire as he spun around. "Come!" He dashed back into the castle, his good friend at his heels.

This Being the End of the True and Accurate Account of the Adventure of

# Howell Davidson

the

# Musketeer

**2021**

Made in United States
Troutdale, OR
11/03/2026

56844426R00218